Success Depends on the Animals

DIANA L. AHMAD

Success Depends on the Animals

Emigrants, Livestock, and Wild Animals on the Overland Trails, 1840–1869

UNIVERSITY OF NEVADA PRESS
Reno & Las Vegas

University of Nevada Press, Reno, Nevada 89557 USA

www.unpress.nevada.edu

Manufactured in the United States of America
Cover image: "Emigrants Crossing the Plains" courtesy of UC Berkeley, Bancroft Library

Library of Congress Cataloging-in-Publication Data
Names: Ahmad, Diana L., 1953-
Title: Success depends on the animals : emigrants, livestock, and wild animals on the Overland Trails, 1840-1869 / Diana L. Ahmad.
Description: Reno : University of Nevada Press, 2016. | Includes bibliographical references and index.
Identifiers: LCCN 2015036824 | ISBN 978-0-87417-997-2 cloth (alkaline paper) ISBN 978-1-943859-10-8 ebook
Subjects: LCSH: Frontier and pioneer life—West (U.S.) | Overland Trails—History. | Overland journeys to the Pacific. | Pioneers—West (U.S.)—History—19th century. | Human-animal relationships—West (U.S.)—History—19th century. | Livestock—West (U.S.)—History—19th century. | Animals—West (U.S.)—History—19th century. | West (U.S.)—History—1848-1860. | West (U.S.)—History—1860-1890.
Classification: LCC F596 .A356 2016 | DDC 978/.02—dc23
LC record available at http://lccn.loc.gov/2015036824

♾ The paper used in this book meets the requirements of American National Standard for Information Sciences—Permanence of Paper for Printed Library Materials, ANSI/NISO Z39.48-1992 (R2002). Binding materials were selected for strength and durability.

FIRST PRINTING
20 19 18 17 16 5 4 3 2 1

In loving memory of my parents,

Allie and Beate

Contents

Preface

BETWEEN THE 1840S AND 1860S, nearly three hundred thousand men, women, and children crossed the American continent to Oregon, California, or Utah looking for a new life in the American West. In the longest voluntary migration of people in history, tens of thousands of horses, mules, oxen, and cattle accompanied the emigrants west. Hundreds of books have been written about the human experiences on the overland trails, while none have been written about the animals that accompanied the emigrants.[1]

During the months of travel, the relationship between many of the emigrants and their animals intensified from one of utility to one of friendship and companionship. The success of the movement west depended in large measure on the animals that hauled the emigrants and their goods. The people, soon to be known as overlanders, moved west during the age of domesticity and changing religious and philosophical views about animals. On the trails many learned firsthand what their ministers and the authors of domestic etiquette literature advocated: be kind to the animals. With that in mind, the travelers learned how to take care of the animals, how to train them, and how to handle dangerous situations.

To the emigrants, understanding where to cross a river, how to pack a wagon, and where to find feed for the animals proved increasingly important. As the animals became worn out from the tasks required of them, the emigrants abandoned them, as they could not wait for the horses, oxen, and mules to recover. Many emigrants tried home remedies or used knowledge gained from contemporary guides to animal health to help cure their animals along the trails. Despite the precautions, thousands of head of livestock died from, for example, malnutrition, alkali poisoning, and snakebite. Some emigrants expressed sadness for the loss of their animals that reflected sentiments beyond the fear of losing the animals' power to pull the wagons. Soon after crossing the Missouri River, emigrants also encountered wild animals, including buffalo, snakes, and prairie dogs. The wild animals fascinated them, even as they sometimes became targets for meals.

Emigrants brought their cultural values and basic knowledge of animals on the trails when they left their homes. Over the miles some emigrants developed relationships with the animals they had not anticipated. Whether the emigrants traveled to seek their fortunes in the gold mines or to start new farms and whether they traveled alone or in groups, many came to value their new relationships with the animals.

Earlier works, including Alfred Crosby's *The Columbian Exchange: Biological and Cultural Consequences of 1492* and Virginia DeJohn Anderson's *Creatures of Empire: How Domestic Animals Transformed Early America*, studied the importance of human-animal interactions, but they focused on the years of the European settlement along the East Coast of North America and the importance of animals during that period. Whereas their studies looked at animals as "agents of historical change," this study considers the more direct interactions of the overlanders with their livestock, as well as with wild animals.[2] It focuses on that relationship in a specific environment, the overland trails, and in a limited era, the mid-nineteenth century. Adding animals to the history of the overland trails broadens the historical experience of humans and animals alike.

This monograph could not have been completed without the assistance of numerous people. The archivists and librarians at, for example, the Bancroft Library (especially Susan Snyder), Huntington Library, Oregon Historical Society, Mormon Library, and California Historical Society provided guidance to collections and offered invaluable suggestions. A University of Missouri Research Board grant and the Love of Learning grant from the international honor society Phi Kappa Phi provided funding for the research that allowed me to complete this work. For their assistance, I am most grateful.

I am also indebted to many friends for their continuing support and encouragement throughout the process of research and writing. In particular, I would like to thank my mentor Susan Flader and Nancy Taube from the University of Missouri–Columbia; Philip J. Johnson, professor of equine internal medicine at the University of Missouri College of Veterinary Medicine; Michael Green from the University of Nevada–Las Vegas; Sondra Cosgrove from the College of Southern Nevada; the staff of the University of Nevada Press; Robin Collier, Melody Lloyd, Cesar Mendoza, and Larry Gragg from the Missouri University of Science and Technology; Norman Crocker; and my family, all of whom offered fine suggestions and encouragement throughout the years. Without a doubt, I also thank Matt Becker from the University of Nevada Press and the readers of my monograph for their help and advice. Finally, I give a nod to my symbolic mentors, the stuffed toy mascots that kept watch at my desk, holding me accountable while I wrote. They represent those whose journeys are recounted in these pages.

Success Depends on the Animals

Chapter One

Going West

A SUCCESSFUL JOURNEY TO THE WEST depended on the livestock that pulled the emigrants' wagons along the overland trails. Aware that the "oxen must be cared for" and that they would be in "sore straits" if they neglected the animals, the emigrants soon realized that the livestock changed how they proceeded west.[1] They read guidebooks, letters from those who preceded them, and articles in newspapers offering advice about the journey. Much of what they read provided suggestions about how to care for the animals, but none of the guidance told the emigrants to anticipate forming a bond with the livestock that they would work with during the three- to four-month adventure. While on the trails emigrants and domestic animals alike became overlanders, together crossing the plains, the mountains, the rivers, and the deserts.

The relationships among many of the emigrants and animals evolved over the months of travel from a pragmatic working association with their horses, mules, and oxen to one of friendship that bonded the travelers and their animals together by struggle. As they moved west the emigrants discovered wild animals unknown to them before the journey. Most of them quickly developed an awareness of the new creatures, while many others developed an appreciation for the wild animals that they had not anticipated. The months on the trails brought the travelers much closer to animals than they had ever been prior to leaving their homes. The animals became more than machines that pulled the plows in their fields at home. They became a vital part of the journey.

Often, emigrants named the animals that pulled their wagons or carried the emigrants' goods on their backs, such as Joel Barnett, who called his horses Charley and Whip and his oxen Old Baldy and Dick. Naming the animals allowed the travelers to form a closer bond with the creatures and likely reminded them of home and those they left behind. Most emigrants understood that the animals "partake of our labors without profiting by them; of our pleasures without enjoyment in them," yet made them "friends even" for the few months that they traveled together as "companions," while still understanding that the animals provided transportation, labor, and revenue.[2] The animals afforded emotional support through the months of insecurity on the trails. The emigrants and the animals shared the experiences of the lack of good water and food, extremes in temperatures, attacks by insects or wild animals, and the difficult crossings of rivers, mountains, and deserts. The relationship that developed between the animals and the emigrants brought some stability to a time when the unknown elements dominated those of the known. It did not take the emigrants long to realize that "every thing [*sic*] depends upon taking good care of your animals" and that "the emigrant is entirely dependent on his team," in order to make it to their destinations.[3]

Starting in the 1840s people from the United States and Europe traveled west with their livestock, seeking new homes and mineral riches. Some emigrants chose California or Oregon because of their fertile fields, land-speculation opportunities, and healthful climate, while others chose Utah to avoid religious persecution. Still others wanted to avoid the Great Plains, believing it to be the Great American Desert. Many people knew about the West Coast because of the mountain men, traders, and missionaries who had traveled and worked in the region during the previous few decades.[4]

Many future emigrants likely read the words of John L. O'Sullivan from the *Democratic Review* in the mid-1840s, proclaiming that it was America's "Manifest Destiny" to spread over the North American continent. O'Sullivan, a Democratic propagandist, scholar, and lawyer, encouraged Americans to expand into Texas and Oregon.[5] Whether the emigrants understood the meaning of the phrase "Manifest Destiny" or not, they expressed similar sentiments in their letters home, writing about the "'march' of impire [*sic*]" to the Pacific, "Yankee pride and Yankee ambition," and the "spirit of enterprise" that moved the people west "for the purpose of settling themselves and [their] families" in the new lands.[6]

The American settlement of the Oregon Country began after the Spanish and Russians ceded their claims to the Northwest by the mid-1820s, leaving the United States and Great Britain to discuss who controlled the lands between the forty-second parallel and the 54°40′ line. In the Joint Occupation Treaty of 1818, the British and the Americans agreed to jointly occupy the area until such time that one nation decided to withdraw from the agreement. Between the treaty and the 1840s, several people pushed for the settlement of the area, including Hall Jackson Kelley, who tried to entice settlers to Oregon with plans to re-create the Massachusetts Bay Colony; Nathaniel Wyeth, who attempted to dominate the fur and fish trade; and also missionaries, such as Marcus and Narcissa Whitman and Henry and Eliza Spalding, who tried to convert the local Cayuse people. Although none of the groups succeeded in their appointed tasks, they spread the word about the value of the area through newspapers and speeches that claimed Oregon had rich lands and a healthful climate.[7] They understood the journey would be difficult, but they "desired *El Dorado* of the West," and they "bade our loved ones adieu, and began the long and tedious journey of two thousand miles, through the wilderness of explored land, inhabited only by wild beasts and savages of the forest."[8]

In June 1846, after the first few years of wagon trains had arrived in Oregon, President James K. Polk signed an agreement with the British, making the region easier to settle as the two nations agreed to extend the forty-ninth parallel nearly to the Pacific Ocean. The agreement gave the United States the formerly disputed area between the Columbia River and the forty-ninth parallel and left the area south of the river, where most of the emigrants had settled, firmly in the hands of the United States. Then in 1850 President Millard Fillmore signed the Donation Land Act, also known as the Oregon Land Law. It allowed settlers to claim unsurveyed lands in the Willamette Valley. Each settler could claim 320 acres, while a married couple could claim a full section of land. The claimants agreed to cultivate the land and live on it for four years, after which they would receive title to the land.[9]

Also during President Polk's term in office, the Treaty of Guadalupe Hidalgo, in February 1848, ceded approximately one million square miles of land to the United States that included California, the Southwest, and Texas. With the territory gained through the treaty with Spain and the acquisition of the Oregon Country, those wishing to move west could do so, still remain in the United States, and not have to worry about dealing with foreign governments. Coincidentally, gold had been discovered in California just a few days prior to the signing of the treaty with Spain. As news of the gold strike filtered eastward, some people believed that God had saved the gold discovery for the time when the United States possessed the land.[10]

While some emigrants waited for the United States to acquire the western lands before moving, others decided to take their chances in California and Oregon without American ownership of the regions. In May 1841 the first wagon train bound for the West Coast left Missouri led by John Bidwell and John Bartleson. This group of approximately seventy people soon joined another group led by Thomas Fitzpatrick, a well-known mountain man, whom missionaries hired to take them to Oregon. The party traveled as one unit until Soda Springs in modern-day Idaho, at which point the group split in two, with the Fitzpatrick group continuing on to Oregon and the Bartleson-Bidwell contingent heading to California on a previously untraveled route. After nearly six months of travel, the Bartleson-Bidwell party became the first group to go overland to California.[11]

Although few moved to Oregon in 1842, emigration in 1843 increased significantly. In that year the first major group of emigrants headed west with nearly nine hundred men, women, and children and several thousand head of cattle, horses, and oxen.[12] During the pre–Gold Rush years of 1840 to 1848, approximately twelve thousand emigrants went to Oregon, nearly three thousand to California, and fourteen thousand to Utah. The totals grew considerably between 1849 and 1860 as a result of the gold discovery in California, bringing the total number of emigrants to more than two hundred thousand to California, more than fifty-three thousand to Oregon, and more than forty-two thousand to Utah.[13] Members of the Church of Jesus Christ of Latter-day Saints (LDS) who traveled to their Zion in Utah generally stayed on the trails on the north side of the Platte River to avoid conflicts with others who traveled in the same direction.[14]

The majority of emigrants came from the Midwest and Upper South, including Missouri, Illinois, Wisconsin, Indiana, Kentucky, and Ohio, while others came from Europe, including England, Ireland, Switzerland, Denmark, and Norway. For many, the move west proved temporary, as they only intended to mine for gold, while others hoped to improve their lives in the new lands. Some served with the US military, and others wanted to meet the Native Americans before their cultures changed with the oncoming American population.[15]

Many of the men, women, and children kept diaries as they moved west, while others wrote reminiscences many years later. Generally, the emigrants wrote much longer diary entries during the first half of the journey, likely because of the excitement and newness of the adventure. As the days became weeks and then months, the length of the entries shortened as the emigrants tired. Some made brief citations, noting only the weather or the miles traversed that day, while others described the day's activities in detail. Both male and female writers ranged between long, flowery descriptions to simple data-only entries in their writings. Children's explanations spent more time on new animals and exciting events, such as stampedes, rather than on mundane comments about the weather. Sometimes emigrants kept diaries in the form of letters home, mailing their writings whenever they reached a place that offered mail services, such as Fort Kearny or Fort Laramie. Most often, the emigrants' diaries and journals started at the Missouri River and ended when they reached their destinations.

Many of the diarists wrote about their animals. Some noted only the number of animals that made up their outfits, the types of animals that pulled their wagons and carried their goods, or that an animal had died. Others described the animals in great detail, noted the animals' personalities and idiosyncrasies, and even wrote poetry about them. Some emigrants mourned the deaths of the animals and sometimes wrote about the livestock using terms of high praise. Many diarists found wild animals exciting to see, especially for the first time. They attempted to identify the new animals and compared them to animals at home, such as prairie dogs to squirrels and antelope to white-tailed deer. Some described animals new to them in a straightforward style, noting that buffalo stampeded or that a rattlesnake slithered away. Whether they wrote about the animals in detail or merely with a few words, many emigrants soon learned that wild and domestic animals became an important part of their lives as they crossed the continent.

No one group dominated writing about their animals. Whether they traveled alone or with family members or strangers, or if they moved west to seek riches or farms, many emigrants described their experiences with and feelings about their animals. To understand the intensity of the feelings for the animals that some emigrants developed, it remains important to recognize the magnitude of the daily tasks and troubles that the emigrants and the animals experienced on the trails.

The emigrants, soon to become overlanders, carried many things with them to the west. They carried the obvious items, such as wagons, tools, and food. They also brought along things that did not contain pounds but bore heavily on their minds, such as religion, philosophy, and the feelings of sentimentality and domesticity. Many also thought that they understood how to train an animal for a task, but they soon realized that they needed to learn much more in order to survive the journey. Despite the real and the imaginary weight, the emigrants quickly learned that success depended on the animals.

Chapter Two

Emigrants Take More than Livestock West

When the emigrants decided to move west, they needed to determine what supplies they required for the journey. Farming equipment, mining tools, and provisions for the travelers, as well as the feed and goods for the livestock, filled the wagons with many hundreds of pounds. In addition, the emigrants carried the burden of the ideas that they held from their various backgrounds and upbringings. These included religious and philosophical concepts and the teachings of the cult of domesticity. They also carried with them the notions presented to them in artworks and the laws regulating the relationships between humans and animals. They did not anticipate the additional weight that these many ideas added to their baggage.

Success on the trails meant that the emigrants needed to alter the preconceptions they held about animals. Most would learn new ideas about domestic and wild animals as they trekked west. Ministers from Protestant and Catholic churches considered animals inferior to humans. Since the seventeenth century, Western religions generally dismissed nonhuman animals and assumed that only humans possessed intellectual complexity.[1] Abrahamic tradition stated that God created humans above all other life forms and provided animals for them to meet their needs of labor and food. While animal sacrifice ended with the rise of Christianity, early Christians believed that humans ranked above animals because the wild kingdom lacked the ability to reason.[2] After the fall of Eve in the Garden of Eden, people considered animals frightening creatures that would attack them. As a result, it became necessary to make animals subservient to humans. Early theologians, such as Origen and Augustine, preached and codified the inferior role animals played in the lives of humans. Christians believed that animals had no souls and that Christ died for humans, not for animals, strengthening the separation between the two groups.[3]

During the seventeenth century, Puritan ministers taught their parishioners that creatures belonged to a howling wilderness. Compounding their teachings, early settlers discovered new animals in North America that added to the uncertainty of human-animal interactions. Others believed that animals served as metaphors for the virtues and vices of humans, such as bulls representing strength and dogs fidelity, while snakes encouraged caution and prudence and wolves greed. Animals also held religious meanings, such as the lamb signifying the sacrifice of Christ and the dove becoming the image of the Holy Spirit.[4]

Churches discouraged having animals as pets because it took time away from worshipping God. Tenderness must be reserved for humans, not animals. Therefore, humans must treat animals kindly but beware of indulging the beasts.[5] Despite that, Evangelical Protestantism encouraged personal self-discipline in an attempt to achieve perfection that would one day result in animals and humans living together happily once more. In the 1840s even children's books advocated that because of Eve's indiscretion, "this sad curse was on the animals too; not by their fault, poor things! But by man's dreadful sin."[6]

Many of the emigrants took away from their ministers the understanding that humans ruled over the animals but that the beasts must be treated well. Despite the role assigned to the animals by theologians, the Bible entrusted humans with the responsibility of caring for the animals, granting the beasts a weekly day of rest and demanding that care be given to injured animals or those that had gone astray (Gen. 1:26; Exod. 23:12; Deut. 5:14, 22:1–4).[7]

The biblical theme of keeping humans superior to animals yet caring for them could be read in *A Natural History of Animals* (1844) by John Bigland, a British schoolmaster. Although originally published in 1828 in England, the American edition of his book came just as people contemplated the move west. The tone of his work modeled the sermons and writings of ministers of the early nineteenth century, emphasizing the importance of treating all animals as the objects of God's "parental solicitude." Heavily religious in tone, he wrote that "wanton cruelty" against animals demonstrated "impiety towards their Creator." Particularly relevant to the future emigrants, he wrote about the horse "as an inestimable gift" from God and that the animals needed to be treated with "attention and kindness" even when the horse became old or injured.[8] While still in their homes preparing for the journey, little did the future overlanders understand the importance of Bigland's words and that they would have to deal with many injured and dying animals along the trails.

While some read about how to take care of animals in books, others heard about it from ministers who came to wish the emigrants a safe journey. In 1853 the Reverend J. H. Avery, pastor of the Congregational church in Austinburg, Ohio, offered parting words to the travelers. His suggestions, like Bigland's, warned those listening not to overdrive their teams; to walk alongside the wagons to lessen the weight the horses, mules, or oxen would have to pull; and to think carefully before leaving "to preclude the necessity" of thinking later. He also encouraged those listening that they should "hope for the best, and prepare for the worst."[9] Avery could only hope that some of them not only listened to but also heard his plea.

Simultaneous to the teachings of Evangelical Protestant ministers to the emigrants on their treatment of animals, the Church of Jesus Christ of Latter-day Saints encouraged its members to take great care of their animals as well. Joseph Smith, founder of the faith, advised his followers that the kinder they treated animals, the greater would be the increase in peace in the world. Brigham Young, who led the Mormons on the overland trails to the Great Salt Lake in 1847 and was a strong advocate of Smith's admonition, entreated people to take good care of their livestock and warned that those who did not would "lay himself liable to censure in the eyes of justice." Young believed that animals could be given the blessings of the priesthood.[10] Additionally, Young rebuked Mormons going west for killing too many buffalo. Knowing that those of the LDS faith moved west to seek new lives, Young's words resounded loudly among the faithful and reinforced the earlier teachings from their non-Mormon ministers.[11]

While ministers taught their parishioners that animals possessed no souls, philosophers studied the possibility that creatures did possess souls, had cognitive abilities, were self-aware, or possessed moral reciprocity.[12] Their ideas predate the birth of Christ. Pythagoras and Aristotle concerned themselves with whether animals have souls. Teaching in the sixth century BCE, Pythagoras believed that all sentient beings, including animals, have souls that can be reborn into another being, even into a different species. Because he advocated compassion for the suffering of animals, he and his followers practiced vegetarianism and condemned animal sacrifice.[13] By the fourth century BCE and the time of Aristotle, views regarding nonhuman animals possessing souls had changed. Now, even plants possessed a soul, albeit the lowest form of one. Animals too possessed souls and could experience physical sensations and emotions. Despite animals and plants having souls, only humans possessed rational souls that allowed them to have abstract thoughts. The value of animals, according to Aristotle, resulted from their ability to provide for humans.[14]

By the seventeenth century, the Enlightenment changed the discussion of whether animals possessed souls and concluded that animals offered little of value to humanity except as a resource. Although animals were not a major topic of philosophical writing during the Enlightenment, writers paid more attention to their value during this era than in previous centuries. In 1637 René Descartes wrote that animals possessed no souls and served humans as little more than machines. With that Descartes and others cut open live dogs to study the animals' physical makeup. In addition to scientific discovery, believing that animals had no souls and could feel no pain allowed people not to feel guilty about eating or killing the creatures.[15]

A century later more philosophers joined the discussion about the significance of animals in the world. Jean-Jacques Rousseau, author of the 1762 *Discourse on Inequality,* found compassion to be a natural sentiment for humans and animals, although humans lost their feelings of compassion as they received more formal education. He claimed that humans could seek perfection while animals could not, but even so animals could still experience pleasure and suffering. David Hume, a mid-eighteenth-century Scottish philosopher, believed humans should be gentle with animals, while German philosopher Immanuel Kant concluded that humans had no moral responsibilities to animals. Concerned more with protecting humans than animals, Kant advocated that abusing animals could lead to mistreating humans. Unlike Descartes, Jeremy Bentham, in his 1789 *An Introduction to the Principles of Morals and Legislation,* believed that animals could suffer and feel pain, although Bentham believed that the animals did not mind working for humans if treated well.[16]

The Enlightenment in the United States also briefly considered animals. In Benjamin Franklin's *Poor Richard's Almanack,* published yearly between 1732 and 1757, saying number 151 proved particularly important to any mid-nineteenth-century people who may have heard about or read his work and contemplated going west. Although written many years before emigration on the overland trails began, Franklin's words fitted the movement west well when he wrote, "For want of a nail the shoe is lost; for want of a shoe, the horse is lost; for want of a horse the rider is lost." Franklin's moral agreed with other Enlightenment philosophers in that he called for humans to be kind to animals. In addition to Franklin, Thomas Paine's *Age of Reason,* written in 1793–94, advocated kindness to animals by claiming that cruelty to them violated the moral duty of humans.[17]

In addition to religious and philosophical ideas, emigrants began their overland journeys in the 1840s during the era of domesticity and sentimentality, when Americans came to believe in the importance of rational self-discipline and benevolence. Those ideas fitted well with the teachings of Evangelical Protestantism that explained that emotionalism showed individual and social progress toward perfection. Gentility and the ethic of kindness became desirable traits in members of the middle class or the Victorian culture in the United States. These ideas led, in theory, to animals being well cared for by their owners. Kindness to animals became a mark of class identity, while cruelty to animals, as Kant stated earlier, served as a predictor of antisocial behavior in adults.[18] Victorians wanted to improve the quality of life and wrote numerous prescriptive books and articles. They offered suggestions about how to raise children and manage the household, as well as gave guidance on learning proper etiquette. Their writings often expressed displeasure with the mistreatment of animals and encouraged their proper care. American Victorians believed that children and animals had much in common, including innocence and goodness.[19]

Between the 1830s and 1850s, publishers released more than one hundred manuals educating readers about manners. School textbook writers included sections on etiquette in the hopes that gaps in home education might be made up in the schools. The most popular included American authors Lydia Maria Child and Lydia Huntley Sigourney and British author Charlotte Elizabeth Tonna. Because of the spread of mass media, their writings could easily be found and acquired. Author of the "Thanksgiving Day" poem, Lydia Child wrote *The Mother's Book* (1831) which addressed itself to American mothers "on whose intelligence and discretion the safety and prosperity of our republic so much depend." She believed her book would address a gap in the literature for the middle class and provide guidance for raising children. Publishing since 1815, Lydia Sigourney's works met with praise claiming that mothers would appreciate her practical and useful advice. Sigourney eventually became the editor of *Godey's Lady's Book*.[20] Charlotte Elizabeth Tonna wrote the 1845 *Kindness to Animals; or, The Sin of Cruelty Exposed and Rebuked* which advocated developing kind relations between humans and animals in an attempt to reconcile the two groups' efforts to re-create the peaceful kingdom on earth.[21]

Coincidentally published during the years surrounding the overland movement, the domestic advice books provided suggestions about how to handle animals properly and carefully. Treating animals with kindness became part of the middle-class ideology. Sigourney warned parents to teach their children about the goodness of animals, such as dogs and cats. She even suggested that children be taught not to step on worms or to destroy ants' houses. Learning to be kind to animals helped not only the creatures; it also helped the children develop into caring, moral adults.[22]

The advice books readily included religious references regarding the care of animals. Since the beginning of the European movement to New England, settlers accepted the idea of people's dominion over animals, and those ideas continued in the advice publications. For example, Tonna's writing reinforced the idea that prior to Eve's disobedience to God in the Garden of Eden, no violence or cruelty existed between humans and animals, but now humans needed to show kindness toward animals, emphasizing that people had a duty to them.[23]

With the need to learn to be kind to animals, people started to keep pets, although only with selected animals, such as cats and dogs. As such, the early nineteenth century saw a growth in the number of animals kept for affection and nurturance. Pet keeping became associated with children because it taught them how to care for animals and develop into loving, caring adults. As a result, Victorian literature quickly incorporated pet keeping into the advice manuals. Parents also needed to become role models for their children and behave properly around animals, as children paid attention to the actions of their family members.[24]

The advice manuals suggested that parents teach children not to be afraid of animals, even the smallest insects. Yet parents needed to explain that some animals, such as spiders, needed to be killed and had the potential to harm people, but to remember that even these threatening creatures came from "the Great Father." In the case where a dangerous animal needed to be killed, the manuals advised parents, and through them the children, that the death must be done "as mercifully as possible." Children also needed to become aware of animals experiencing "distress" and to learn how to help them by watching how their parents handled the various situations. In part, that meant that children needed to understand that their duty to animals included protecting defenseless creatures.[25] As the emigrants traveled west, they discovered animals such as rattlesnakes, wolves, and mosquitoes that could hurt them as well as their livestock. In many cases, how they treated the dangerous creatures reflected their Victorian-era training.

Protecting animals also meant that children must not be permitted to inflict injury on them, including teasing them or causing the animals to be fearful. The manuals taught children to correct animals when the creatures had done something wrong, but to beware of beating the animals "too severely" as that might cause the animals to rebel against their owners. The lessons taught the children that abuse of animals also insulted God and therefore should be avoided. During the 1830s–1850s, newspapers reinforced the idea of kindness to animals and reported cases of animal abuse.[26]

Particularly relevant to people who traveled the overland route, advice manuals taught them to have "love for all innocent creatures." This advice likely encouraged some to have a more open mind about the animals they encountered. The manuals encouraged people to train animals to do their best by praising and rewarding them in order for the creatures to become attached to the humans and to do as the people desired. Of particular importance to the emigrants, advice manuals considered horses the "most sensible and most affectionate of creatures," but noted that horses experienced "a great deal of dreadful pain" when forced to pull heavy carriages. In 1848 Catherine E. Beecher, author of *A Treatise on Domestic Economy,* explained that women should know how to properly care for domestic animals and provided some suggestions on how to do so. She said that horses needed to be shod every two months if they worked a lot, fish oil and strong perfumes should be applied to them to keep away flies, and horses should not be fed or watered until they had cooled off from their labor.[27] Advice writers such as Beecher, Child, and Sigourney could not have known that the emigrants would meet many animals new to them as well as work with domestic livestock along the trails. Their suggestions likely caused many animals to be treated more humanely than they might otherwise have been.

In addition to advice manuals, the middle and elite classes in the Victorian period developed an interest in art that included the natural world. During the early republican period, Americans demanded portraits of national heroes and statesmen, but by the 1830s artists began responding to requests for paintings of rural life and animals. American artists had a popular appeal, and attendance at galleries in cities such as New York increased dramatically.[28] Trained in Europe and in the United States at such schools as the National Academy of Design and the American Academy of Art, American artists by the 1830s no longer needed the patronage of the elite to finance their profession as the middle class developed an interest in their work and purchased their paintings. With that, animal portraits became a part of the Victorian interior decoration that demonstrated the gentility of that socioeconomic group. As rural families gained financial status, they purchased pictures of cows that they often hung in their dining rooms, while paintings of horses hung in tack rooms or other areas dominated by men.[29]

The art of the era included landscape paintings that illustrated wilderness scenes and showed the glory of God. To depict animals, American art expanded on the ideas of the "howling wilderness" and the "peaceable kingdom." In the howling wilderness artists portrayed animals as frightening and waiting to be subdued by humans, while the peaceable kingdom showed humans and animals living happily next to one another, reflecting the ideas in the Old Testament book of Isaiah. The notion of a howling wilderness descended from the European medieval bestiaries. Bestiaries linked animals with human virtues and vices. An American bestiary developed in the colonial era that included animals new to the colonists. Like its European counterpart, the American bestiary combined scientific treatise with morality tales and included real and mythical animals. With many immigrants coming from Great Britain, they undoubtedly brought with them a fear of wolves because of the myths and stories they had heard since childhood, despite the fact that wolves had been eradicated in their country since the sixteenth century.[30] The nightly howling of the gray wolves and coyotes on the Great Plains must have reminded them of the stories they had heard in their youth.

In other works of art, animals took precedence in landscapes instead of human civilization. Founded by Thomas Cole, the Hudson River School stressed the pastoral and the domestic and often included animals in the scenes. The artists, including Cole, Asher B. Durand, and Albert Bierstadt, used animals to evoke moods, such as tranquillity, solitude, power, and majesty. George Catlin, another member of the Hudson River School, urged protection for animals in his 1841 *Letters and Notes on the Manners, Customs, and Condition of the North American Indians.* Prior to these painters' work, American painting rarely included animals, except for children shown with pets.[31]

Many painters began their careers as portrait or itinerant artists who traveled around the rural United States seeking clients. These artists often included animals in their scenes of New England and the Adirondack Mountains. They painted barnyard scenes and even barns as financial necessity demanded. One of the most significant barnyard artists was Thomas Hewes Hinckley, who in 1845 received a commission to paint Daniel Webster's Ayrshire herd. His work for Webster led to further sales of his paintings to the American Art Union, including *The Rabbit Hunter* in 1850. He continued to paint scenes with animals in them, and the Royal Academy of London and the Centennial Exhibition in Philadelphia displayed his works during the 1870s.[32]

A number of other American artists during the early to mid-nineteenth century focused on birds. The finest included Alexander Wilson, a Scottish immigrant who drew birds clearly and with precision. From 1808 to 1814 he published the nine-volume *American Ornithology* which illustrated 268 species of birds, including some that had not been described earlier. Following Wilson's work, the paintings of John James Audubon, an immigrant from France, superseded those of Wilson with the publication of *The Birds of America* (1827–38). Although not published with the average person in mind, the color-plate book contained more than a thousand life-size illustrations of 500 species, including more than two dozen new species.[33]

As Americans developed a greater interest in animals from their ministers, the advice manuals, and the paintings, some sought to protect the creatures by a change in the animals' legal status. Legal condemnation of animal abuse began in mid-seventeenth-century Massachusetts. In 1641 the General Court of Massachusetts passed the Body of Liberties which contained two sections on the treatment of animals. Although not a strong statement on animal protection, it still afforded the animals a modicum of safety. Liberties 92 and 93 resulted from the desire to protect animals because abuse against them violated such human concepts as discipline and virtue. As domestic animals belonged to people, outlawing cruelty protected human property while saving the animals from potential abuse.[34]

One hundred years later, the British began to protest blood sports, such as cockfights and bullbaiting, which eventually led to the 1822 English law that protected certain animals from mistreatment. Then in 1824 the British organized the Society for the Prevention of Cruelty to Animals. Influenced by the British action, the states of New York (1828) and Massachusetts (1835) passed laws defining gratuitous cruelty against horses, cattle, and sheep and made such violations a misdemeanor punishable by a small fine or a few days in jail or both. The laws applied only to animals owned by residents and did not apply to strays. By 1866 fourteen states and six territories possessed laws calling for the punishment of those who abused domestic animals. Of those areas three states

(Missouri, Iowa, and Oregon) and four territories (Kansas, Washington, Nevada, and Idaho) lay along the overland trails, making the passage of the laws likely influenced by the emigrants' experiences. The growing interest in the care of animals led to the creation of the American Society for the Prevention of Cruelty to Animals in 1866. Laws against cruelty to animals soon also included the federal 1873 "Twenty-Eight-Hour Law" which required rest, food, and water for animals every twenty-eight hours on the animals' way to the slaughterhouse.[35]

Philosophy, religion, domesticity, art, and law appear to be varied and diverse topics, and certainly do not carry any actual weight, like flour, horseshoes, and feed, yet for the emigrants, the future overlanders, the ideas and actions of those beliefs, ideas, and conceptions shaped their notions about the animals that carried them west. Emigrants did not cite the philosophers or discuss the art they may have seen, yet most of them understood the need to care for the horses, mules, and oxen that carried the items that weighed hundreds of pounds. Messages from the advice manuals and the ministers likely influenced them more directly. The advice and examples the emigrants received from the sources called for people to restrain themselves vis-à-vis the animals, in particular domestic animals. The training the emigrants received taught them self-discipline for the benefit of themselves as well as their animals.

Some of the overland emigrants developed a relationship with their livestock and the wild animals that they encountered on the journey. Those who did so treated the animals, wild and domestic, in a way that would have made ministers, philosophers, and writers of advice manuals proud. The overlanders knew that in order to succeed in their journey, they needed to take care of their animals. Many of them went far beyond the basic needs of providing food and water. Some emigrants developed an emotional bond with the animals that helped them to journey across the continent. Yet as in other relationships, once the need for the animals ended, the relationship between the humans and the animals changed. But for the three to four months on the trails, some emigrants experienced relationships with the animals unlike anything they expected or anticipated when they left their homes. The relationships they developed with their animals meant the success of their ultimate aspirations in the West. To achieve that success, they needed to learn the best animals to take, what supplies to purchase, and how to work with the animals. The animals would be with them around the clock for several months, and the emigrants' relationships with those animals, domestic and wild, needed to be positive to ensure that they arrived at their destinations.

Chapter Three

Preparation for the Journey West

Preparations for the journey west required as much attention to the needs of the oxen, horses, and mules as to the needs of the human travelers. Planning and training for the adventure began long before the emigrants arrived at one of the jumping-off towns, yet once they reached the communities of, for example, Independence or St. Joseph, Missouri; Kanesville, Iowa; or Winter Quarters, Nebraska, their need to complete final arrangements reached a fever pitch. These communities provided the last known places to purchase supplies, draft animals, and obtain other items required for the trip. Here, the travelers learned how to ride horses or mules, how to hitch the oxen to the wagons, and what supplies the animals required. The Missouri River border towns provided their last chance to prepare for the long journey ahead.

Coming from the United States and Europe, the travelers often headed to specific towns because of recommendations from agents representing the communities. The agents understood that the soon-to-be overlanders needed to purchase last-minute items and even whole outfits before heading west. The seasonal profits from this brisk business fostered the economic success of the towns. Agents often misrepresented the costs, availability of items, and conditions in communities that vied with those of the towns that the agents represented.[1] City newspapers printed "For California" columns on their first and second pages during March, April, and May, advertising items and providing information to the emigrants in the hope of luring the people to their communities. The newspaper editors provided testimonials about the advantages of their towns' businesses that included endorsements from such luminaries as General Stephen W. Kearny, Commodore Robert F. Stockton, and Joseph Meek, lending a sense of trustworthiness to the newspapers' claims about their towns. The newspapers claimed that the roads in their communities serviced the needs of the travelers better and more efficiently than other roads. Emigrants also served as promoters for one community over another through their letters home, describing where travelers could obtain more favorable rates or informing that diseases, such as cholera, could be avoided by going to a different place.[2]

Emigrants often arrived at St. Louis via the Ohio and Mississippi Rivers. If they traveled, for example, from New York, they had already journeyed approximately one thousand miles, with nearly two thousand more to go. Traveling up the Missouri River on steamboats saved the livestock several hundred miles of pulling heavily laden wagons. In the 1840s more than two thousand vessels carried people, livestock, and supplies to Independence, St. Joseph, and points farther up the river, while in the 1850s steamboat traffic increased to more than three thousand vessels. Often crowded with passengers, livestock, and outfits for the journey, the steamboats could be heard for several miles before arriving at the docks.[3] By 1859 travelers could take a train to St. Joseph and then board a steamboat to communities in Kansas and Iowa. Between 1846 and 1863 more than half of the emigrants traveled by train or steamboat to St. Joseph, saving the labor of their animals for later.[4]

Independence, Missouri, became the first major jumping-off town for the trails bound for Oregon and California. By the early 1840s Independence already possessed a reputation as a community devoted to travelers heading southwest on the Santa Fe Trail. Established in 1827 after the town of Franklin, Missouri, washed away in a flood, Independence became the most important upriver community for western emigration through 1847. Since the 1820s Independence businessmen had outfitted traders who traveled to Santa Fe, New Mexico, and as such developed a good sense of the needs of overlanders. The Independence merchants also supplied the mountain men, including William Sublette, who trapped and traded for furs in the Rocky Mountain region.[5] In 1843 the first major Oregon-bound group left Independence with approximately nine hundred emigrants and more than eighteen hundred domestic animals.[6] Soon, tents and wagons filled with emigrants bound for the West occupied the surrounding community.

By 1847 Independence began to experience competition for overland traffic with the communities of St. Joseph, Missouri, and Kanesville, Iowa (renamed Council Bluffs in 1853). Located closer to the Platte River than Independence, St. Joseph, called St. Joe by many, became the jumping-off town of choice for overlanders, particularly if their point of origin included Ohio, Indiana, Illinois, Iowa, or Michigan. Established in 1827 by Joseph Robidoux as a trading post, the community was organized by the state of Missouri as a town in 1845. A few months later the town licensed ferry facilities on the Missouri River that serviced the needs of merchants as well as the thousands of emigrants who passed through the area.[7] As a result of the movement west, the population of the town nearly quintupled, from 682 in 1845 to more than 3,000 in the 1850s. Farmers who surrounded St. Joseph profited by the sale of produce to the overlanders which in turn allowed the farmers to pay off their land debts. In addition to farm products, the value of the stock of goods sold in the town reached more than four hundred thousand dollars in 1849. Calling St. Joseph a "new and flourishing Town beautifully situated" on the Missouri River, the overlanders credited the success of the community to the economic contributions made by those on their way west.[8]

Still farther up the Missouri River, Kanesville, Iowa, developed in 1846 when members of the Church of Jesus Christ of Latter-day Saints stopped there on their way to the Great Salt Lake area of Utah. Considered a more advantageous starting point than St. Joseph or Independence, Kanesville offered better ferry service than the other communities and had less illness, especially during the 1850 cholera outbreak along the overland trail. After 1851 approximately 60 percent of the overlanders left from this Iowa jumping-off town.[9] Although Independence, St. Joseph, and Kanesville experienced the largest number of overland emigrants, other communities also served as departure cities, including Westport–Kansas City. Native Americans from the Sac, Fox, Delaware, and Wyandot Nations visited the community and sometimes sold "little shaggy ponies" to the emigrants. Westport entrepreneurs sold entire outfits to travelers even before the travelers left their homes via mail-order services.[10]

These towns served as places for emigrants to gather and prepare for the long journey to their destinations. Some travelers stayed for only a few days, while others stayed several months. Hotels rented rooms but did not have enough space for the vast number of emigrants who arrived in their communities. As a result travelers sometimes rented cabins or stayed with the towns' residents. The countrysides surrounding the communities quickly filled with tents, wagons, and livestock and resembled towns themselves.[11] Camping allowed the emigrants to get accustomed to living in the confined space in and surrounding their wagons. Living outside the towns permitted them to train their animals, grow accustomed to them and their behaviors, as well as gain the trust of the livestock. While they organized for the journey, tended to last-minute purchases and repairs, and waited for the grasses to sprout on the Great Plains that would provide sustenance for the animals on the trails, many emigrants attended lectures and religious services at Catholic, Baptist, and Presbyterian churches or visited gambling and drinking establishments that provided "all kinds of wickedness."[12]

Local businesses advertised that the travelers could find anything the emigrants and their animals needed at their stores. The emigrants took advantage of the dry goods stores, bakeries, clothing shops, boot and shoemakers, and gunsmith shops; for their livestock they purchased items at harness and saddler stores, ox-yoke makers, blacksmith businesses, and livery stables.

Agents in St. Louis "egregiously imposed on" emigrants, trying to convince them that the border communities' prices far exceeded those in their city. Buying the goods in St. Louis also meant that the items needed to be shipped or hauled to the jumping-off towns, increasing the cost of the purchases accordingly. Those who waited to buy items along the Missouri River found the prices to be "as cheap as any other" community and in sufficient supply to furnish the needs of the emigrants.[13] Ultimately, the cost of items in St. Louis, St. Joseph, and other communities depended on the abilities of the purchasers to negotiate a desirable price for the goods as well as the demand for the items.

One of the most important purchases travelers made included "good and substantial wagons" with covers, as the journey could take several months. The wagons needed to be strong and light and made with well-seasoned timber. They needed to be easy for the oxen, horses, or mules to pull. As rivers and streams would need to be forded, guidebooks recommended corking the wagon bed so the wagon could float while keeping the goods inside dry as well. The cost of the wagons ranged from twenty-two to ninety dollars and varied with the wagon's quality and size.[14] The supplies needed for human use on the trails included such things as a tent, blankets, guns, rifles, pistols, knives, cooking utensils, and table furniture. Food items for the emigrants included flour, cornmeal, hard bread, crackers, bacon, ham, tea, coffee, sugar, dried fruits, beans, and rice. To prevent scurvy, emigrants brought along horseradish and potatoes prepared with apple vinegar. These items could be purchased in the jumping-off towns.[15]

The needs of the livestock that would pull the wagons westward proved as vital to a successful journey as the items the emigrants required for themselves. As such, the travelers purchased numerous things for the care and well-being of the horses, oxen, and mules. The total cost of items for emigrants who came from urban areas exceeded that incurred by farming families, as the city people needed to purchase practically everything for the journey, whereas the others already possessed many of the supplies needed for the trip.[16] The animals' needs included yokes, bows, pickets, packing saddles, halters, bridles, hames, whiffletrees, and sixty-foot hemp ropes.[17]

Emigrants purchased shoes and nails for the animals with the thought that "many a valuable ox was left on the road the last season [1851], for the want of a SINGLE SHOE," likely reminding some of Benjamin Franklin's saying in *Poor Richard's Almanack*.[18] Numerous blacksmith forges operated in the jumping-off communities that offered shoeing services for the animals. Independence, in 1847, had forty-seven blacksmith forges and employed four to five hundred men serving the needs of the emigrants and their animals, while in Kanesville, William F. Carter offered his blacksmithing services "on the shortest notice" for those passing through the Iowa town. The "incessant hammering and banging" at the blacksmith shops resounded throughout the communities. Businesses also offered boarding and stabling services for the livestock.[19]

For those who needed to purchase all of their animals or those who wished only to supplement or replace livestock, the jumping-off towns offered many horses, oxen, and mules for sale. St. Joseph and Independence maintained good reputations as places to acquire animals at reasonable rates. Merchants smartly brought horses, mules, and oxen to their communities to sell to the emigrants. In anticipation of acquiring stock before too many others arrived at the respective communities, emigrants sometimes sent members of their party ahead to get the best animals available. If emigrants purchased animals in St. Louis, they needed to consider how much wear and tear the livestock would experience on the trail from the Mississippi to the Missouri River. With that thought some ferried their animals up the Missouri River to their jumping-off town, but that added cost made buying the livestock along the Missouri River a logical decision, much to the financial advantage of the upriver merchants.[20]

At St. Joseph, as in other communities along the Missouri River, men rushed "to and fro, all eager to buy mules, horses, and oxen for the journey." Despite the demand for the animals, even during the Gold Rush in 1849, the cost of purchasing the livestock varied little throughout the 1840s and 1850s. Often sold in pairs, or by the yoke, oxen at the major outfitting centers of Independence and St. Joseph averaged forty dollars each. During the same years mules sold for an average of sixty-six dollars, while horses sold for sixty.[21] If funds allowed for extra purchases of livestock, a few emigrants bought a horse and carriage for a spouse traveling with them or an extra horse or mule that looked promising. Others bought younger and lighter stock to replace the teams they acquired earlier or those they had brought from home that had already walked hundreds of miles.[22] Horses, oxen, and mules sold or exchanged by their emigrant owners could be rested, or recruited, and then resold by local merchants to unwary emigrants in the market for a fast purchase. The ability of the buyers to negotiate likely played a role in the final price the travelers paid. The importance of acquiring good animals could not be overstated. John Hawkins Clark, in 1852, summed it up well when he wrote that "good horses, good mules and good oxen are everything on a journey like this."[23]

The cost of buying an ox, mule, or horse involved more than the purchase price of the animal. The emigrants needed to decide upon the animals that would be best for the challenge of the cross-continent journey. Overwhelmingly, emigrants chose oxen over horses or mules to pull their wagons. John D. Unruh Jr. and Merrill J. Mattes, the premier historians of the overland trails, claimed that oxen made up 50–75 percent of the draft animals.[24] The exact number of livestock on the trails remains unknown. Emigrants usually preferred one animal over another, but sometimes they had to settle for their second or third choice because of the lack of availability of the desired animals or their limited finances.

Oxen, slower but stronger than horses and mules, traveled at one-third to one-half the speed of horses, making the journey approximately three weeks longer than with other animals. Considered "as safe, if not THE SAFEST animals that can be employed upon the road," oxen were believed to be easier to drive than other animals and less likely to stampede or be stolen by Native Americans, although the heat could be hard on them. Additional attributes of the ox included its being considered noble, durable, and gentle.[25] These castrated adult male cattle made excellent "pullers" of the wagons and possessed calm and patient personalities. Many emigrants tried to obtain four yokes of oxen, or eight animals, per wagon. With luck they had enough money to buy five or six yokes, allowing them to rotate those pulling the wagons while keeping one yoke free, permitting those oxen time to walk along with the team without needing to work.[26] Emigrants preferred the oxen to be four to six years of age. Oxen that died along the trail provided fresh meat that could be consumed by the emigrants or sold to those around them, providing a break from the salted meats the travelers regularly ate as well as earning some much-needed cash.[27]

Some emigrants brought cows on the overland journey as well. Although "less serviceable for the yoke," cows substituted for oxen when necessary. More important, cows provided milk. With the milk put into a jar with a lid, the movement of the wagon churned it into butter and buttermilk as the wagons rolled forward. While passing through desert areas, the milk provided something to drink for the emigrants. Cows stopped producing, however, when the trails passed through alkali areas "so strongly impregnated with poisonous matter as to be dangerous for drinking purposes."[28]

After oxen, some emigrants preferred mules to pull their goods over the trails. They considered mules quicker than oxen and hardier than horses. Created by humanity, mules cannot reproduce. Mules result from the crossbreeding of male donkeys (jacks) and female horses, while female donkeys (jennys) and male horses produce hinnies.[29] Mules have existed for at least three thousand years, and the people of Asia Minor, Greece, and Rome used them as pack animals. Homer's *Iliad* and the Christian Bible cite the use of mules, and many centuries later Napoléon used the animals while going through the European Alps.[30] In the United States, in the 1780s, George Washington became interested in breeding mules and received jacks and jennets (small Spanish horses) from the king of Spain and the Marquis de Lafayette. In 1827 Henry Clay, the Kentucky statesman, imported Spanish jacks to be bred with his state's mares.[31]

Starting in the 1820s the mule came to Missouri from the Southwest rather than from the East as part of the Santa Fe trade with Mexico. In 1823 a group of Santa Fe traders returned to Missouri with four hundred jacks, jennys, and mules, and by 1832 more than one thousand animals came from Mexico.[32] Emigrants received their introduction to the hardy animals when they arrived at the jumping-off towns. Known to be stubborn and ill-tempered, mules soon won the favor of some emigrants in spite of their reputation. Mules were called "long-eared animules" or "long-eared horses," and the emigrants decided that they would not be traveling in as good a style as with horses, though they believed that the mules would provide good service on the way west. At first the emigrants who preferred mules expressed shyness toward the animals, but after working with them found the negative comments they had heard exaggerated. In 1849, at the start of his journey, Niles Searls described his mules as a "venerable, sedate looking chap" with his "youthful companion," demonstrating an open mind to the new animals in his life.[33] The emigrants had many months and miles to decide the accuracy of their first impressions of the animals.

Unlike oxen or mules, horses had a reputation for being a pleasant team if chosen carefully. Omnipresent in the United States, horses worked with people on farms and in cities, providing the power needed for transportation and farming in the settled areas of the country.[34] Yet when moving across the continent, few recommended using "the common horse" for the journey. Instead, some who had traveled to California and Oregon previously stressed that "no American horse can be taken entirely through" to the West Coast and that the "shabbiest Shawnee pony" would provide better service to the emigrants.[35]

Once the travelers acquired the best animals available, they purchased food for the livestock meant to be consumed during the early days of the journey. Although the emigrants hoped that the grasses along the trails would support the livestock, they understood that in April and early May the grasses had not yet grown long enough to feed the animals. In addition, the emigrants prepared for the possibility of prairie fires and rain flooding parts of the trails, reducing the naturally occurring grasses.[36] Unlike the cost of livestock, the price of corn and hay varied a great deal at the river towns. Between 1849 and 1853 prices ranged from 25 cents to $2.00 per bushel for corn. Emigrants paid "all prices" for hay, but generally it cost between 30 cents and $1.75 per one hundred pounds. Farmers in Iowa and Missouri quickly learned how to raise prices as the demand for their products rose, and they "were not bashful" about taking advantage of the emigrants.[37]

During May 1856, in anticipation of the lack of grass along the trail and understanding the need to take care of his animals, Finley McDiarmid from Wisconsin purchased, in spite of the costs, an extra wagon specifically for hauling grain for his animals. He figured that by the time the livestock consumed the supply, the grasses along the trail would have grown sufficiently long enough to allow the animals to feed on the available natural supplies. After the animals consumed the purchased grains, McDiarmid abandoned the wagon, freeing up the animals that pulled it and giving him extra teams in case he needed them. Henry Packer from Illinois, who also bought extra food supplies for his livestock, warned that those who did not do so would "suffer the consequences of their folly."[38]

With horses, mules, oxen, and feed purchased, it came time for the emigrants to begin training the animals, as well as themselves, for the journey ahead. Not all of the emigrants knew how to hitch the livestock to a wagon. Some of them had never been on an animal prior to their arrival at the river communities. The training needed to be efficient and as good as possible. Most of the emigrants quickly learned that spending time with the animals proved advantageous. The majority of the animals had never been broken, and they needed to become accustomed to human touch and to spoken commands. "Noise and commotion" best described the scenes along the Missouri River.[39]

Many emigrants hired men to break the animals and to teach the travelers how to handle the livestock. The training took place near the campgrounds of the travelers. The emigrants found the teamsters' abilities to handle fifteen- to twenty-foot whips to "wallop the leather" of the oxen "extremely amusing," but understood that they needed to pay close attention to how the men performed their tasks, as the emigrants would soon be the ones doing them. Often, Mexicans and black Missourians hired themselves out to teach the emigrants how to the break mules to the saddle or oxen to the yoke. The men charged five dollars to break the animals, "although they did not perform their duty very thoroughly," according to Amos Batchelder in 1849. Undoubtedly, Batchelder changed his view about the fee his trainer charged, admitting that mule breaking proved "hard upon both mules and men." In the "science of mule breaking," trainers taught the emigrants to catch mules with a lasso and to pack the animals with sacks of sand as a training device. They learned that the tasks needed to be performed daily for the benefit of man and mule.[40]

Training the animals and men took much longer than anticipated and involved numerous dangerous incidents. Mules caused the most trouble for the emigrants, as the animal's "natural disposition is not the most amiable, and his disposition not to go the way you want him to is proverbial." Famous for laying down, rolling over, breaking their saddles, and other "ungainly capers," mules often delayed the starting dates for numerous emigrants.[41] Oxen also caused some emigrants trouble during the days of training, such as in 1851, when Charles A. Brandt encountered problems with his oxen while removing the yoke from one member of the team. The other half of the team then galloped away, swinging the yoke in all directions in an effort to get the yoke off its neck. Without achieving success at removing the encumbrance, the young ox became submissive and allowed Brandt to release the yoke without difficulty.[42]

Although mules held the worst reputation for their lack of cooperation with the emigrants, horses too rebelled when put into harnesses and kicked and bucked until freed from their duties. With the failure to train some of the animals successfully, sometimes emigrants faced the necessity of trading the troublesome livestock for other animals, especially oxen, in the hopes of better success. Acquiring whatever animals remained available, emigrants sometimes traded for animals of lesser value, such as replacing a mule with a pony or a "little Mexican horse."[43] Through perseverance and time, the training of the animals worked, and the livestock learned to carry the saddles, wear the yokes, and pull the loads given to them by their new masters.

At the same time the animals learned their new tasks, the emigrants learned how to manage them. New to the job at hand, many emigrants believed that those who observed their attempts to train the animals laughed at their efforts to steer three or four yoke of oxen in a straight line or at being bucked off an untamed mule or horse. Coming from urban areas, many of the emigrants had never seen an ox prior to arriving in Missouri, and others learned how to ride a mule or horse in St. Joseph, Kanesville, or Independence. They admitted to shyness at the start of the training sessions and that their sense of humor often failed them as the animals refused to do as the emigrants desired. Within a week or two, the training became less arduous and fearsome, as they developed the needed skills for handling their stock.[44] Likely, many of the emigrants remembered the lessons about being kind to animals in the books they had read and in the sermons they had heard.

Published guidebooks, letters from friends and relatives, and newspapers provided tips for training the animals. Dozens of guidebooks existed that emigrants purchased for less than one dollar. The authors had sometimes traveled the routes themselves or acquired their information from others. They offered recommendations about the trails to take, the equipment to carry, and the livestock to choose for the journey. The best guidebooks included William Clayton's *The Latter-day Saints' Emigrants' Guide* (1848), Hosea B. Horn's *Horn's Overland Guide to California* (1853), and Randolph Marcy's *The Prairie Traveler* (1859).[45] The authors hoped their advice provided practical and up-to-date information. In J. M. Shively's *Route and Distances to Oregon and California* (1846), he claimed that his "only design is, that you be fitted out in a proper manner, and have your pilot in your pocket." The authors often had ulterior motives for writing the books that included promoting areas in the West for political reasons, hoping to sell land in California or Oregon, or wanting to make money from the sale of the guidebooks.[46]

Regarding animals, a few of the authors emphatically recommended that "the greatest attention should be paid to all animals that are taken on this journey" and that the success of the journey "depends entirely" on the condition of the teams.[47] Other advice provided practical suggestions, such as telling the readers to avoid old campgrounds as they could be unhealthy for the stock, not to hurry the animals along the trail, and to recruit the animals regularly.[48] The guidebooks emphasized not burdening the teams by overloading the wagons with too much "trumpery." Pointing out that only necessary items should be taken west, they highlighted that for the livestock, "the labor shall be only that of TRAVELING, and not of HAULING" the goods and people.[49] In 1853 J. Tarbell's guidebook ended with "Drive slow; the old adage is applicable, 'haste makes waste.' Keep cool and let your moderation be the predominating principle, and you are ensured a safe but tedious journey."[50] The decision to abide by the advice would be left to the travelers.

As the emigrants consulted their guidebooks and made last-minute preparations, a heightened sense of "bustle and confusion" emerged throughout the towns. Animals could be found everywhere, tied to trees, running at large, or standing quietly by their wagons. In the evenings the mules brayed, making "the forest resound with their unearthly music." The thousands of people and animals made "quite a crowd" and led William Braden in 1849 to conclude that even mules counted in the numbers of emigrants bound for the West.[51]

Last-minute preparations included finalizing the groups with whom they would travel, who would take charge of the animals, and what they would do with the dogs they brought from home. For millennia, dogs had accompanied people on their journeys, and traveling to the West proved no exception, as early-nineteenth-century explorers, including Meriwether Lewis and William Clark, Wilson Price Hunt, and Peter Skene Odgen, brought dogs on their expeditions. Despite the successful history of bringing dogs west, a few early emigrant groups, such as Lansford Hastings's 1842 and Solomon Tetherow's 1845 companies, feared that the dogs' barking and running among the animals might stampede the livestock. They also worried that the dry conditions on the trail might cause rabies in the dogs. Originally, Hastings's group issued a "dog decree" that called for "the immediate and the indiscriminate extermination of the whole canine race" for the safety of the wagon train. After killing a few of the dogs in Hastings's group, a number of the owners of the remaining dogs announced that whoever killed their dogs would face the same fate as their canines. That announcement resulted in the overturn of Hastings's antidog decree. Although not as drastic as the Hastings rule, the bylaws of Tetherow's Savannah Oregon Emigrating Company permitted its captain to shoot any dog running loose around the camp.[52] The journals of the group did not note if the captain shot any of the company's dogs.

In other emigrant groups arguments in favor of dogs becoming a part of the company included using canines as "sentinels" that would warn the emigrants about the presence of Native Americans and that they would provide a greater sense of security to the wagon train. Dogs provided companionship, worked as hunters, and, if necessary, could be eaten. In other cases, owners of dogs chose to leave their pets at home with relatives. The dogs howled as their masters pulled away from the family home, and some refused to eat again and died.[53] The concern about the canines' lives and welfare, as well as the threats made to those who shot a few of the dogs in Hastings's group, demonstrated that the emigrants cared for their pets. The journey would determine if they would extend their feelings to other animals in their group, as well as to the wild ones they met.

With the wagons prepared, the animals trained, and the supplies purchased, crossing the "big Mosura" came next. At the major crossing points, such as Kanesville, St. Joseph, and Weston, every type of team and wagon filled the banks where the emigrants registered their wagons and livestock and purchased tickets that gave them access to the steamboats that would ferry them to the western side of the river.[54] The cost of the ferries ranged between four and ten dollars per wagon and ox team. In 1846 the town of St. Joseph set ferry rates and imposed fines for operating unlicensed ferries.[55] Yet with so many thousands of animals and wagons crossing the river, regulating the fees and ferries proved difficult. When they purchased their tickets, the emigrants received a numbered place in line and crossed when they heard their number called. Several hours to nearly a week passed before the emigrants heard their numbers, as the demand for ferry transportation exceeded the number of vessels available, although some emigrants, wagons, and animals crossed earlier than scheduled by using "cunning management."[56]

Getting the livestock onboard the ferries proved to be one of the many challenges for some emigrants as they set out on their journey. Keeping the animals on the ferries proved to be another. In 1852 John Hawkins Clark drove one of his mules on board a ferry at St. Joseph, and just as soon as he turned around to get the next mule, the first one jumped into the river. Clark's group then decided to drive all the mules on board at the same time, succeeding admirably. Once onboard, all the animals moved simultaneously to the same side of the vessel, tipping the boat into the water and frightening the mules, who proceeded to leap overboard. Once more, Clark and his companions gathered the mules together and drove them on the ferry, this time successfully. From there they drove the oxen onboard and after a few more challenges finally "got the horned brutes" to the western side of the river. Those onshore cheered and applauded the efforts of Clark's group as they, no doubt, worried about how their animals would react when their turn came to load the ferries and to cross the river. As for Clark's response, he wrote that "there was no swearing done, for nobody could be found that could do justice to the occasion."[57] How the emigrants and livestock would respond in the future to the dozens of rivers they would have to cross would likely be another story.

Once safely across, the emigrants and the livestock set out on their three- to four-month journey, with some of them taking the trails to Oregon, some to California, and others to Salt Lake City. Many variations of the routes existed, but at first the majority of the emigrants followed the Platte River through Nebraska, journeying past Fort Kearny, Ash Hollow, Chimney Rock, and Scotts Bluff. From there they started through Wyoming, where they passed Fort Laramie, Independence Rock, Devil's Gate, and South Pass. Just after South Pass the trail split, with those heading to California and the Great Salt Lake turning southwest to

Fort Bridger and on to Salt Lake City. The Mormons stayed in the City of the Saints, while the California-bound emigrants continued on through Nevada, the Humboldt Desert, and over the Sierra Nevada into California. Those bound for Oregon turned northwest after South Pass and traveled to Soda Springs and Fort Hall, then along the Snake River into the Cascade Mountains, and eventually down the Columbia River.[58]

Purchasing and training the animals proved to be just the beginning of the relationship between the emigrants and the livestock as they prepared to cross the continent. These preparations proved to be vital to the success of the journey west. Training the animals taught the travelers that they needed to learn how to work with the livestock to benefit all those concerned. The better care given to the animals, the quicker and more successful the journey would be.

The journey, in many ways, relied on the animals. For some, the livestock and the emigrants formed a mutually beneficial partnership. In effect, the animals trained the emigrants at the same time that the emigrants trained the animals. The travelers arrived at the jumping-off towns with ideas about animals formed by the teachings of the era. The preparations for the journey showed the organizational skills of the emigrants, but the impact of the journey for many of them taught them that their relationships with the animals far exceeded anything they had previously considered. Many emigrants came to see the animals as cotravelers, not just as the haulers of goods. An unanticipated, intangible relationship formed over the next months between some of the humans and the livestock that none had anticipated.

The travelers purchased the best horses, mules, and oxen that they could afford. They hoped that the journey would be as trouble free as possible. In 1852 Lodisa Frizzell, on the other hand, feared that the animals "will share the fate of the 'gallant grey' of Snowdouns Knight," referencing Sir Walter Scott's narrative poem "The Lady of the Lake" (1810), where the exhausted horse died in a hunt by an impatient rider.[59] The days to come would tell the tale of how the emigrants, soon to become overlanders, treated their animals, how the travelers and the animals worked together, and what made the journey successful.

Chapter Four

On the Trails

Once across the Missouri River, the needs of the livestock ruled the lives of the emigrants. Now came the time to see if the training the two sets of travelers received in the jumping-off towns worked. Many emigrants possessed guidebooks that told them the trails to follow and the likely places to find water and grass. They quickly learned the significance of the landmarks they passed, giving them a sense of how far they had traveled. The best routes provided suitable campsites, easy river crossings, and smooth trails, but as the years went on the routes deteriorated due to the thousands of wagons, people, and animals that journeyed west. Easily accessible grass and water disappeared, forcing the emigrants many miles off the roads for their evening campsites. Most important, the emigrants quickly learned that the animals dominated the needs of the groups, and they came to understand that the livestock required a great deal of attention.

Crossing rivers safely, negotiating favorable ferry rates, and handling the animals in all types of weather challenged the emigrants throughout the journey. Animals that died or disappeared during the night needed to be replaced. The emigrants learned how to handle natural, as well as man-made, problems. Although they did not know it at the time, the first several weeks on the trails would be their easiest. The worst parts of the journey came near the end, when the animals and the people had the least amount of energy and resources available. The time had come to make the needs of the animals an even greater priority in order for the emigrants to arrive at their destinations safely.

Many of the lessons the emigrants mastered remained the same during the main years of overland emigration. From the first days of travel, many learned how to encourage their animals to move forward, pulling the wagons filled with thousands of pounds of goods. Guidebooks and advice from friends and relatives recommended breaking in the animals slowly and driving only short distances at first. The snapping of the ox whip, whistling "I'm Bound for the Happy Land of Canaan," and knowing an "inexhaustible vocabulary of cuss words" encouraged many animals to move along the trails. In 1853 Hannah Cornaby, a member of the Latter-day Saints, disagreed with the third incentive, explaining that the oxen did not understand English, and as such curse words would have been lost on the animals and did just as well without them. During the first few days after leaving the Missouri River, the livestock "decidedly objected to being caught, as if they suspected what was before them," but soon the animals became tamer and gentler and accustomed to the daily routines.[1]

At the start of the continental crossing, most emigrants allowed the stock to become acquainted with one another and to travel slowly, yet deliberately, until the animals became "broke in" to the routine of the trails. Breaking the animals in at a steady pace with short and easy drives proved essential to the success of the journey. It trained the animals to understand the labor the emigrants expected of them. Starting early in the morning, around dawn, the wagon trains pulled out. They traveled until approximately midday, when they stopped for a "nooning," where they rested the animals, as well as themselves, during the heat of the day. Whenever possible, the emigrants rested near grass and water to allow the animals to graze, yet as the years went on the emigrants needed to take their livestock several miles off the trails to find suitable feed. Keeping the livestock at a steady walk, sixteen to eighteen miles could be achieved on a good day. If conditions permitted, the animals needed to be allowed to drink morning, noon, and night.[2]

After the first few long days of travel, the emigrants, now becoming overlanders, learned the journey required vigilance, patience, and energy. Even the nights gave the emigrants no break from paying attention to their animals. Upon selecting a suitable stopping point for the evening, the emigrants drove their wagons into a large circle, forming a corral, and then took the animals to a watering and grazing spot. Then just before turning in for the night, the emigrants drove the animals into the wagon circle and connected the lead and rear wagons with chains, essentially locking in the animals until morning and keeping them safe from theft by Native Americans or from wandering too far from camp. The fear of Native Americans stealing the cattle and horses even led to the recommendation that horses needed to be tied to the wagon wheels during the night to provide an extra layer of protection. As the overlanders settled in until morning, the night guards kept watch over the encampment and the animals.[3]

Emigrants often picketed or staked mules and horses in areas with grass. Picketing consisted of taking a twenty-to-thirty-foot rope, tying one end of it to a sharpened stick approximately twenty inches in length and tying the other end around the animal's neck, and then driving the stick into the ground, providing the animal a forty-to-sixty-foot circle in which to graze while being unable to run away. During the first few days of the journey, picketing animals near one another also allowed the animals to become acquainted with one another, in addition to making sure they did not stray.[4] Occasionally, while being picketed an animal became tangled up in the rope, tripped, and broke its neck. The nearby wild animals, such as coyotes, benefited from the loss of the animal.[5]

In time and after the animals learned the daily routines, emigrants turned the animals loose near the wagons during the evenings. As when the emigrants corralled the livestock in the wagon circle during their early days of travel, they maintained watch over the animals to hinder theft and to prevent the animals from wandering too far away as they searched for better grasses. In the morning the guards drove the livestock to the campsite, where the emigrants caught, harnessed, yoked, or saddled the mules, oxen, or horses for the day's travel.[6] The guards learned to pay attention to the behavior of the animals during the night, as horses and mules sensed the approach of strangers, be it animal or human.[7]

After traveling on the trails for a week, emigrants camped on Sunday, "partly to observe the Sabbath and partly to let our teams rest." No doubt remembering the teachings of their ministers and the books by such writers of the 1840s as Charlotte Elizabeth Tonna, who reminded readers that Sundays should be kept to rest animals, most emigrants incorporated Sundays as official rest days into their wagon train company's bylaws. Phoebe Judson, in 1853, declared that "I'm glad that He did not forget the poor, tired animals, and said, 'Thou shalt not work thy ox.'" She went on to declare that the only places along the trails that made her happy were those that provided "pure water and good grass for our cattle, and allowed them to rest over the Sabbath." Judson believed that her group lost only two of the fifty livestock they started with because they rested their animals every Sunday.[8]

The overlanders acknowledged that the animals pulled heavily loaded wagons, and, "in justice" to the livestock, they hoped for no Sunday travel "unless it was actually necessary." Claiming that "Safety was our Motto," Newton Finley's group made it a point to rest regularly on Sundays.[9] The Sabbath did not mean a day without labor, as most emigrants spent the day taking care of their livestock, as well as themselves. They shoed horses and oxen, oiled the harnesses, and repaired the wagons, as well as took care of their own needs, such as washing and mending clothing.[10] Occasionally, overlanders found it necessary to travel on Sunday or during the night in order to find water or feed. At other times, if they found good resting sites on Saturday, they chose to "keep Sabbath with the Hebrews."[11]

The availability of grass and water dictated the resting places of the overlanders. As the years progressed and the number of animals, people, and wagons that crossed the continent increased, the opportunities for locating good grass and water for the livestock near the trails declined. Conditions quickly worsened west of South Pass just as the overlanders made the decision to go northwest to Oregon or southwest to California or Salt Lake.

As the emigrants wound their way along the trails to South Pass, they found reassurance in passing landmarks, such as Ash Hollow, Courthouse Rock, and Devil's Gate. They heard about these geographic markers from relatives, friends, and guidebooks. In 1853 Phoebe Judson saw the landmarks as "mile stones [*sic*] along the journey of life," and passing each one made the trip that much shorter.[12]

On the Platte River trail, as well as other trails to come, locating good water and grass for the livestock became a motivating force for the overlanders. They possessed "little time or inclination for romancing" about the lands they crossed, claiming that their attention must be concentrated upon the needs of the animals. Once they learned how to spot potential water and grass spots, they could "comfortably and philosophically enjoy the trip" while simultaneously keeping an eye into the distance for likely places to rest. Often, those distances seemed deceptively close, but in reality proved to be many miles farther along the trail.[13]

Those who left from Independence or St. Joseph first encountered the Big Blue and Little Blue Rivers in eastern Kansas, while those who departed Kanesville and Council Bluffs first crossed the Elkhorn River. No matter their jumping-off community, they headed toward the Platte River which would guide them for hundreds of miles. There, the emigrants "felt ourselves fairly launched out 'on the plains'" and finally on their way west.[14] The banks of the Platte, a wide, warm, and shallow river, consisted of "huge piles of sand" "destitute" of vegetation, but covered with prickly pear, whose thorns caused lameness in the animals.[15] As the years wore on, the deterioration along the Platte in part resulted from the intensive grazing by the emigrants' animals and the earlier travel on the route. Despite the dwindling land resources, the Platte contained "good and sweet water," while grass and sometimes timber grew on the small islands in the river.[16]

Many overlanders blamed flooding, fires, and buffalo for the lack of grass along the Platte, often without acknowledging their animals' roles in its deterioration. During the spring, melting snows in the western uplands wound their way down the rivers and flooded campsites, inundating the grassy areas and making them unavailable for the stock and causing emigrants to hitch their horses and drive out as quickly as possible, even during the overnight hours. Prairie fires burned the grass, leading A. A. Enos, who traveled in 1850, to express pleasure about "the wisdom of hauling our load of snapped corn from Kanesville" to feed the animals.[17] Without considering buffalo indigenous animals to the Great Plains, the overlanders blamed the lack of grass on the bison that "devoured" it, causing the land to appear like "an over fed [*sic*] pasture in a dry time." As the miles westward accumulated and the number of buffalo declined, the amount of vegetation for the domestic animals increased, relieving some of the worry the emigrants felt for their stock. To lessen the demand for the available feed, emigrants, if possible, traveled in smaller groups, resulting in lower population pressure on the grasses.[18] By the end of the 1850s, even with fewer emigrants moving west, many of the trails offered less and less nourishment for the animals, no matter where the wagons traveled.

The trails from the Missouri River towns merged at the Platte near Fort Kearny. Established in 1848 and on the eve of the California Gold Rush, the fort protected and aided the overlanders as they passed through the area. To ensure that the US Army's livestock had sufficient grass to eat, the government reserved lands within one mile of the fort for only military animals, as the army could not risk losing the available feed. As many overlanders approached Fort Kearny, they claimed that their "most and greatest anxiety is to get good feed for our teams; it is upon their ability to perform the journey that we all rely." They also noticed that "our beards seem to grow faster than the grass on the prairies" near the fort.[19] Perhaps because of the limits placed on their animals' grazing by the military, as well as by the lack of grasses, they realized that little could help them except taking as many precautions for their animals as possible.

Fort Kearny had a blacksmith shop available for use by the emigrants. If the wagon train companies had a blacksmith among the overlanders, the military allowed them to use the shop at no charge. The military kept tallies of the number of men, women, and stock that passed the fort, although not all travelers stopped there and as such were not counted. At the fort, described as a "neat little place" with "very pretty residences," the government kept approximately a hundred "splendid" horses and as many mules for use by the 150 troops stationed there.[20]

Traveling on both sides of the Platte before and after the fort, livestock and wagons lined the trails "as far as we can see" and made "quite a lively scene." As the wagon trains moved forward, the distances between the groups slowly widened, creating, almost unbelievably, times of silence, causing J. D. Breyfogle in 1849 to remark that "not a single sound greets my ears, save now and then the twitter of some wild bird." Despite the growing distances between the wagons, the number of animals remained nearly the same, and the feed sources continued to deteriorate because of the demands made upon the grasses by the livestock. Most often passing Fort Kearny in May, the overlanders understood that the grasses would be better in June, yet they knew that the cattle, horses, and mules passing the fort would eat the grass "as fast as it grew" no matter the month they traveled through the area.[21]

After the military installation, the overlanders continued along the Platte and found that the river remained wide, shallow, and with a quicksand bottom. The situation for the animals along the river route remained the same, with little acceptable grass but good water, until they reached Ash Hollow, shortly after the Platte split into the North and South Platte Rivers. Following the north fork, they entered the river valley, where they found readily available wood and grass as well as good water. Filled with ash trees, the region provided the hilliest part of the trail since leaving the Missouri River. The emigrants praised the area, stating that "nothing we have yet seen can exceed the beauty of Ash Hollow," and noted the thousands of birds, clear stream, and plentiful wood supplies, claiming that the scene "will never be effaced" from their memories.[22] Undoubtedly, most emigrants saw the area as a great relief for the animals as well as themselves.

Conditions improved for the emigrants and livestock for the sixty miles between Ash Hollow and Courthouse Rock, another of the "milestones" along the journey. This landmark resembled a "large public edifice" with a jail-shaped rock nearby. On this stretch of trail, the grass improved, and C. W. Smith commented, in 1850, that the "grass [is] better than we have before seen," and "in fact, this is the earliest [May 1850] period at which the grass can be considered fit for working cattle."[23] Inevitably, the many days of seeing poor grasses and no trees made Ash Hollow more beautiful than they would have thought at the start of their journey.

A dozen miles west of Courthouse Rock and seventy-five miles west of Ash Hollow, overlanders came upon another of the many "natural curiosities" along the westward trails, Chimney Rock. Emigrants found that this geologic formation looked like a conical hill with a smokestack on top. Chimney Rock could be seen from a distance of twenty to thirty miles. If the emigrants camped close enough to the formation which rose nearly three hundred feet into the air, they could pay a "moonlight visit to this curious freak of nature."[24] Only a few miles west of Chimney Rock, the emigrant trains arrived at Scotts Bluff, named for Hiram Scott, a member of William H. Ashley's company of mountain men, whose traveling companions left him for dead near his namesake bluffs in the 1820s. Here, wild sage could be gathered and hay made for the teams.[25] For the horses, oxen, and mules, Chimney Rock and Scotts Bluff proved to be just another stopping point on the trail where conditions for them remained largely unchanged.

Next, the overlanders arrived at Fort Laramie, located at the confluence of the Platte and Laramie Rivers in eastern Wyoming. Established in 1834 by William Sublette and Robert Campbell as Fort William and built of sun-dried bricks and surrounded by fifteen-foot walls that provided a narrow palisade, the American Fur Company absorbed the fort two years later. In 1849 the US military purchased it with the intention of providing protection for the overlanders. Later, the military used the installation as a base for their dealings with the Sioux and eventually abandoned the fort in 1890.[26] Impressions of the fort varied widely, even during the same few weeks in 1853, as William R. Brown called the fort a "beautiful place" and Dr. Thomas Flint found the facility "all in dilapidated condition" only three weeks later. As at Fort Kearny, the military kept tallies of the emigrants and the livestock that passed through the area.[27]

From those who had preceded them, as well as their guidebooks, many overlanders knew the trail between Forts Kearny and Laramie provided poor feed for the livestock. Because of this information, many purchased corn at Fort Kearny to supplement the meager grazing along the trail. By the time they reached Fort Laramie, they began to supplement the animals' grazing and corn ration with one quart of flour a day. The flour provided little sustenance; however, it made the livestock feel full. The overlanders experienced the scarcity of feed for their livestock and feared for those who followed them, noting that "what those behind us will do is more than I can say."[28] By this point on the journey, many emigrants understood the impact their animals, their fellow travelers, and their wagons had on the environment they passed through.

After leaving Fort Laramie in June 1850, Dr. James S. Shepherd stated that "there is not grass enough on an acre to keep a sheep from starvation." Others concurred with the doctor's observations, finding the grass short and dry and the soil poor and sandy. Occasionally, the overlanders had to travel nearly a day's journey of twelve miles to find grass along the fifty-mile stretch past Fort Laramie. Despite that, sometimes luck prevailed, as when the captain of James Cowden's group discovered a small valley approximately a half mile off the main road and surrounded by high rocky walls. With an entrance only one animal wide, Cowden's group spent the night there without the fear of needing to share the grasses they had found with others. For those less fortunate than Cowden's group, they managed to find good water in the La Bonte River, even though the grasses remained of poor quality.[29] Sometimes, only a few days made a difference between the amount of grass one overlander's animals found versus another person's animals, and each traveler hoped that his or her livestock found the grass first, as "feed is getting scarcer all the time."[30] The emigrants maintained a constant vigilance for feed for their livestock, understood the importance of being among the earliest along the trail, and took time to choose a good spot for the night. Picking a campsite depended in great measure on the availability of something to eat for the animals, and the overlanders often had to take the animals away from camp for several miles before finding suitable grasses for them to graze upon during the night.

Along the trail just west of Fort Laramie, emigrants entered into the Black Hills of Wyoming and could see the often snow-covered Laramie Peak in the mountains to the south. This part of the trail had a "perpetual succession of ups and downs" and a reddish clay soil with sharp stones that "cruelly crippled the feet of the oxen." The animals limped along as the barren hills provided no sustenance for them. Phoebe Judson's husband "drove over the stones as carefully as though they were eggs," hoping to avoid injury to their stock.[31] Finding suitable water sources for the animals proved as difficult as good grazing sites in the area west of Fort Laramie and the Black Hills. Emigrants found numerous skeletons scattered about as "evidences of its baneful character," with the water described as resembling "the contents of a dye pot."[32] Feed, water, and trail conditions determined a successful day's travel, with most emigrants taking it one day at a time.

Continuing along the North Platte River to Mormon Ferry, near present-day Casper, Wyoming, they headed toward Willow Springs and on to Independence Rock and Devil's Gate. At Willow Springs, despite its name, they found conditions worse than previously, with the grass "about the same as of Main-street, Racine [Wisconsin]." The willow trees, "struggling for an existence in a sandy desert, amid the gray sage and thorny cactus," provided no feed for the animals. That resulted in their being "much reduced in strength" due to the poor grazing for the last several hundred miles. The overlanders started to notice that the number of dead livestock began to increase along this stretch of trail, even finding eight dead cattle in a single pile.[33]

As the travelers transitioned their route to follow the Sweetwater River, a tributary of the North Platte, they encountered roads "as smooth and as hard as any McAdamised [*sic*] road" in the East. Despite the smoother conditions of the trail, little vegetation existed, except for the trees that shaded the travelers when they rested. From the campsites, the animals needed to walk another 6 to 15 miles to find sufficient grazing.[34]

Once moving along the Sweetwater River route, the overlanders came upon several landmarks, one practical and the others curious. Approximately 1.5 miles before Independence Rock, the emigrants discovered Saleratus Lake. Here, William Clayton's *The Latter-day Saints' Emigrants' Guide* (1848) instructed his readers to "gather your Saleraetus [*sic*]" from the lake. The water on the lake contained a four-inch crust of sodium bicarbonate or baking soda "of the finest quality" that could be used as a leavening agent in baking.[35]

An important landmark for the overlanders, Independence Rock earned its name when William Sublette celebrated July 4, 1824, at that location. At 136 feet high, 2,000 feet long, 700 feet wide, and approximately 1 mile in circumference, Independence Rock looked like "an enormous elephant kneeling down." On this outgrowth thousands of travelers engraved their names and the dates of their visit. Some overlanders, such as Eliza Ann McAuley, found names of friends who preceded her family to the West, while others, like Maria Belshaw, knew none of the people who had carved, painted, or used tar to add their names to the rock.[36] If the overlanders managed to be at Independence Rock on July 4, they celebrated the birthday of the United States with singing and dancing and a sense that "we must have been watched over and protected by an invisible army of the old Revolutionary soldiers."[37] For the animals, the Independence Rock region provided "first rate grass" and good water, permitting the stock to satisfy themselves, despite the thousands of horses, oxen, and mules that passed through this region over the years.[38]

Approximately 5 miles past Independence Rock, the Sweetwater River cut a 360-foot-high, narrow gorge with perpendicular rock walls, known as Devil's Gate. As it was impassable by wagons, the overlanders passed to the south side with their goods and animals. Many emigrants climbed the walls of this "curiosity worthy of a traveler's notice" and etched their names there just as many had done at Independence Rock.[39] The areas after Devil's Gate provided "excellent pasture" because of the "numerous springs" along the trail. They also found Ice Springs located just past Devil's Gate, where the emigrants discovered ice by digging into the ground between 6 and 36 inches. Some thought the ice had a salty taste, while others made lemonade.[40]

Another 120 miles along the Sweetwater brought the overlanders to a gradual ascent to South Pass (7,412 feet elevation), with some of them not even noticing the increase in altitude because of its gentle slope. South Pass separates the Wind River Range to the north and the Great Divide Basin to the south. Here, they crossed the Continental Divide, where they "bid adieu to the waters, which make their way into the Atalantic [*sic*], & now we drink of the waters which flow into the Pacific."[41] About 15 miles farther, the emigrants arrived at Sublette's Cutoff, where they made the decision to either take the Cutoff and travel through 50 waterless miles or take the longer, though well-watered, route to Fort Bridger. At this point, the end goal of the journey, California or Oregon, largely determined the decision to take one cutoff or the other, and that decision impacted their animals.

After the turnoff for Sublette's Cutoff, emigrants encountered the Little Sandy and Big Sandy Rivers. In this region overlanders needed to dig into the sand to get a "little brackish water" and drive their stock 3 to 4 miles up hillsides to find feed. Near the Big Sandy the spring contained sulfur, and the stock refused to drink it.[42] From that point until the Green River to the southwest, "a kind of desert" existed that the overlanders and their stock needed to cross. The emigrants gathered as much grass as they could before starting out across the waterless area. Few found little more than one or two small sacks of grass even after a half day of cutting. Joseph E. Ware's guidebook warned readers that the miles between the Big Sandy and Green River contained "not a drop of water," but that if the emigrants traveled during the coolest hours of the day and night the stock would not need it. Unfortunately, Ware's guidebook proved inaccurate by 15 miles, as 50, not 35, miles separated the two rivers. Feeling deceived by Ware's guide, Dr. James S. Shepherd admonished the readers of his letters home to prepare for a "full fifty miles without water."[43]

Many looked forward to reaching the Green River, where they hoped the grasses and water would refresh their animals. Instead, the grassy plain that they anticipated consisted only of "thorny cactus and bitter greasewood," and the overlanders could not witness their cattle "luxuriating on the succulent grass and slaking their thirst with the crystal waters."[44] Because of the lack of feed, the emigrants turned their animals loose after crossing the Green River and left them without guards because the stock needed to range widely to find something to eat. The animals found only wild sage, greasewood, and "an occasional bunch of grass being the entire product of the soil."[45] After the Green River, conditions for the animals improved, and soon the emigrants found "first-rate" grass in sufficient quantities to permit the animals to feast while they rested.[46]

As the emigrants moved along the trails, the weather challenged them as much as obtaining good water and feed. Thunder, hail, and snowstorms frightened the animals. Some overlanders reported hail ranging in size from bullets and walnuts to hens' eggs and peaches striking their animals and causing them pain.[47] To prevent the horses and mules from running away, the overlanders held the animals or tied them to the wagons. Lightning strikes, sometimes only yards from the encampment, caused oxen to fall to their knees and put pressure on the head, chest, and arms of the emigrants.[48] Mules refused to move during storms but "turned tail to the wind" or, "in the language of the seaman, scud before the gale." Calling the storms "part of the spices of this journey," emigrants could do little but hold onto their animals.[49] While lightning and hailstorms provided variety, daytime temperatures could reach the triple digits on the Great Plains, and the Sierra Nevada's nighttime temperatures could freeze oxen to death. After the snow and rain, the animals looked like "drowned rats, shivering from the cold," if they survived.[50]

Controlling horses and mules proved easier than restraining cattle and oxen during the storms. Thunderstorms often caused the livestock to stampede, and if the overlanders corralled the animals in the wagon circle, overturned wagons resulted from the pressure of the cattle breaking through the enclosure. The cattle ran to get away from the storm, and "nothing could be done to secure them." With luck, the oxen and cattle would be found shortly after the storm.[51]

Stampedes resulted from more than thunderstorms and proved to be more "dreaded upon the plains than almost any disaster that can happen." Dogs running about the camp, children jumping in front of a team, horses unfamiliar with another group's animals, or things that only the livestock understood caused stampedes. Horses and mules frightened easily during the night hours, and some guidebooks recommended preparing "several good horses" to chase after those that had taken flight. Often over quickly, stampedes overturned wagons, ran over other livestock, and injured or killed overlanders. Dead oxen or cattle provided meals for the next day.[52]

Luckily, stampeding animals rarely ran too far away, and overlanders quickly rounded them up after the animals stopped running. During the nights livestock often wandered off from camp as they grazed away from the wagons. Gathering the animals in the morning cost many emigrants "the best traveling part of the day," and sometimes the animals could not be found, forcing the overlanders to travel with fewer than the desired number of livestock pulling the wagons.[53] While looking for missing animals, overlanders found stock belonging to emigrants unknown to the finders. Stray oxen, mules, or horses could be incorporated into the finders' teams, replacing lame animals or adding to the livestock accompanying the emigrants. In the search for the missing stock, overlanders sometimes lost their way, and parties needed to be sent to look for the missing men.[54] Emigrants learned that they needed to keep a lookout for the animals while simultaneously keeping track of the wagons' location, especially if the train proceeded on the day's journey.

Just like emigrants, some livestock "did not like the emigrating business" and "like *sensible beings*" took the homeward route. In 1849, as John Benson entered Kansas, he encountered three men returning to their homes in the East because they missed their families, but he also encountered an ox and a mule returning East, "no doubt for the same reason." The two animals walked as close together as if they had been yoked and did not respond to overlanders' attempts to get their attention, as "they no doubt, had something serious on their minds."[55] Emigrants quickly learned to watch out for the animals that "appeared to have had enough of this wonderland" and attempted to return home every evening. Regularly, they captured the cows and horses and returned them to the night's campsite.[56] Sometimes, oxen and cattle joined herds of buffalo, while overlanders' horses joined herds belonging to Native Americans.[57] Likely, the overlanders put their own apprehensions about the journey onto the livestock and wished for the freedom to be able to return home to what they and the escaped animals considered an easier life in the East.

In addition to animals wandering off during the night, returning home, or being acquired by others, many oxen, mules, and horses became lame or died along the trails due to exhaustion and poor water and feed. These animals needed to be replaced in order to complete the journey as efficiently as possible. Trading posts along, for example, the Platte and Bear Rivers and in the Humboldt Meadows, run by retired mountain men, the French, Native Americans, Mormons, and others, bought lame stock, rested them, and resold them to another overlander. Traders generally paid $3 for a lame heifer or steer, although rates went as high as $18 per head.[58]

The cost of purchasing replacement animals on the trails proved expensive and depended on how late in the season the overlanders started. Emigrants often purchased cows and steers to serve as alternates for their teams as the original team members' health deteriorated. Acquiring new animals permitted the task of pulling the wagons to be shared among the livestock, thereby affording some relief to the most seasoned members of the team. The cost of horses, ponies, and mules ranged from $25 to $200 each at the various trading posts.[59] A yoke of oxen commonly sold for $45–$60 along the Platte, yet traveling late in 1851 Jean Rio Pearce's group paid $465. The price her group paid likely reflected supply and demand, as they had earlier purchased an ox for $30 at Fort Kearny.[60]

Native Americans served as other suppliers of livestock for the overlanders. Experiences with the Native bands along the trails varied between romantic images of Native Americans chasing wild horses to Indians stealing the animals belonging to the emigrants. The overlanders encountered numerous indigenous groups, including the Pawnee, Sioux, and Shoshone. From other travelers some overlanders heard that Indians crawled among the animals at night while wearing a wolf skin or other disguise to steal the horses, but many reported that they experienced little trouble with the local Native groups.[61] As the overlanders familiarized themselves with the techniques of traveling, they soon learned how to protect their stock by setting up nightly watches.

For some Native Americans, stealing the emigrants' animals became a business, as they sold the horses or cows back to the overlanders for cash or goods. Indians sold horses for $30–$50 a head, the going rate for the animals.[62] As with the traders along the trails, lame or sick animals could be traded or sold to the Indians for fresh stock. Then the Native Americans nursed the animals back to health and sold them for higher rates to other overlanders. As the sellers preferred, emigrants paid for the animals with provisions instead of cash. These exchanges included pills for "congestive fever," brown sugar, clothing, and rifles. While the overlanders negotiated with the Indians for the animals, the emigrants also purchased fresh produce, such as potatoes and peas, from them.[63] In an extreme case, in 1859, near Salt Lake City, a group of Indians from "noted tribes" offered to exchange their ponies for a group of Mormon women and girls. The male members of this group of Latter-day Saints declined the offer and moved on along the trail.[64]

Traveling for many weeks, the emigrants negotiated for new animals, located feed and water, and guarded their animals. Passing landmarks became a routine. The trails became more challenging as the miles continued into the deserts and mountains, while supplies dwindled. As the overlanders approached the turnoff for California or Oregon, few realized what lay ahead and that for the animals the last miles would prove to be the worst part of the journey.

Now facing mountains and deserts and with eight hundred to nine hundred miles to go, many overlanders lightened the loads in their wagons once again to save the worn-out livestock as much as they could. Ever since crossing the Missouri River, emigrants regularly abandoned hundreds of thousands of dollars' worth of goods in an effort to lessen the burden on the animals. To save weight they repacked items in sacks and discarded the heavy trunks, barrels, and boxes.

They also cut off parts of the wagons to save even more weight that the livestock needed to pull. Many of the emigrants understood that they needed to lighten the loads to make the pulling easier for the animals. Simultaneously, other emigrants thoughtlessly threw away hundreds of pounds of corn for the horses, perhaps believing that the grasses on the plains would be sufficient.[65] At Fort Kearny emigrants sold provisions at a "great sacrifice," but three hundred miles west, by Fort Laramie, thousands of dollars' worth of goods, such as sheet-iron stoves and silverware, lay abandoned on the sides of the trail as the overlanders realized that "many of these things [are] about as much use to the emigrant as two tails to a dog."[66] By the time the overlanders reached Fort Laramie, the strain on the animals had become apparent, and they realized that they must lighten the wagons' loads even more.[67] The emigrants brought things they believed they would need at their destinations, but they overestimated the strength of their horses, oxen, and mules and underestimated the challenges of the trails.

While some of the overlanders bound for Oregon turned west at Sublette's Cutoff, others continued southwest to Fort Bridger, on the Black Fork of the Green River. There, the remaining Oregon travelers turned northwest, while those headed to California continued in a southwesterly direction. In 1843 James Bridger and Louis Vasquez established Fort Bridger as a trading post and stood ready to "accommodate the pilgrim" with whatever needs he or she might have. The Mormons took over the fort in 1855 before it became a US Army post in 1858.[68] The fort served as yet another of the mile markers on the journey.

From Fort Bridger to the Bear River, wild flax provided excellent feed for the stock, but the descent into the Bear River valley proved challenging. The trail wound around the mountainsides with a "roaring stream" five hundred feet below. In 1849 Amos Batchelder's team experienced the dangers of the trail when his horse, tethered to a mule, lost its foothold and sent both animals down the side of the mountain. Luckily, the animals ended up in the bushes, survived unhurt, and returned to descending the mountain along the trail. Once in the valley the fragrant grasses "of the most superior quality" allowed the livestock to wade "in wild oats to their eyes." The few days along the Bear River provided the "best roads and the best watered of any Days drive" since the emigrants left home.[69]

As with the trail between the Missouri River and Fort Bridger, the emigrants encountered interesting sites after the Bear River region. In particular, Soda Springs provided another natural curiosity worth seeing, with its water resembling a "Sody fountain." The livestock and the overlanders drank the water that made "a fine effervescing drink."[70] Other springs in the area, such as Beer Springs and Steamboat Springs, fascinated the travelers as well with the bubbling waters that "corks would not hold" and "puffed like a steamboat." In addition to the fascination with the nearly blood-warm waters, emigrants took water with them as none existed for the next twenty miles.[71]

Another sixty miles brought the overlanders to Fort Hall on the Snake River. Fort Hall was established in 1834 by Nathaniel Wyeth and was the first permanent US post west of the Continental Divide. Wyeth sold out to Hudson's Bay Company in 1837. They later abandoned it in the mid-1850s. During the post's years of operation, emigrants often traded horses for flour or cattle for horses there. Conditions in the area proved harsh, as by mid-July grass grew only a few inches and soon withered, leaving the overlanders little choice but to serve flour to their animals to supplement their diets.[72]

Between Fort Hall and the Raft River, another turning-off point for those who decided to go to California, overlanders wrote few comments in their diaries, as exhaustion likely deprived them of the energy to record their thoughts during the evenings. In their reduced comments, as before, they noted the poor quality of grazing available in the area. Once again, they lightened the wagons, eliminating the remaining items, including rocking chairs and trunks.[73] At Three Island Crossing the trail split, with some emigrants continuing to follow the Snake River, while others followed a trail a number of miles north of the Snake. Either route they took, the overlanders met again just west of Fort Boise, built in 1834 as another part of Hudson's Bay Company operations.[74] At the fort the Snake River turned north and brought the Oregon-bound emigrants nearer the Blue Mountains.

By the time the emigrants reached the area west of Fort Boise, they had become adept at crossing rivers with their livestock and wagons. Since leaving the Missouri River many weeks earlier, they had crossed innumerable streams and waterways. The knowledge gained during those crossings provided them important experience as they made their way to the Blue Mountains and on to the Columbia River. Crossing the rivers required training, patience, and skills developed over many attempts. Rivers and streams provided needed refreshment for the animals, but the waterways also proved to be dangerous for all involved.

Early on the journey methods of crossing rivers, such as at the Big Blue or the Loup Fork of the Platte, involved complicated maneuvers, including dugout canoes to carry the overlanders' possessions and several days to complete the crossing after the emigrants thoroughly investigated the river's banks and its conditions before attempting to cross.[75] As they followed the Platte they learned to cross the mile-and-a-half-wide river diagonally and not to stop on the crossing for fear of quicksand. They also learned that double-teaming, sometimes up to eight yokes of oxen, worked well to pull the wagons across the river. At the rivers without quicksand, they attempted to make the crossing in stages by taking one team of livestock to the opposite side of the river, securing them, and then returning to the original side to bring a wagon pulled by another team. Once the wagon made it to the middle of the river, the first team's rope would be tied to the wagon's tongue, and the two teams pulled the wagon the rest of the way across the river. This proved to be a tedious and dangerous task.[76]

Taking the animals across the waterways often involved a one-on-one with the stock. To encourage all the animals in the group to cross the river, emigrants tied a lariat to a lead animal and swam the horse or mule to the opposite side in the hopes that the remaining animals would follow the lead. Or emigrants rode a lead animal into the river in an attempt to encourage the rest to follow, and soon those crossing could hear "the languages from all nations where [*sic*] here dealt out with many blows and more curses to the poor animal. The *women* taking an *active warm part* in the *game*."[77]

Despite precautions, emigrants and livestock drowned on a regular basis while attempting to cross rivers or because the emigrants attempted to save their animals. The victims often became tangled up in the ropes meant to guide them along the safest route. Deaths also resulted when the animals missed the area where the riverbank had been cut down for them to allow easier access to the water, and the river current carried the livestock away, drowning them in the process. Fear led the animals to run over their teammates, as well as emigrants who attempted to help them, resulting in death for the animals and the overlanders alike.[78] Crossing rivers caused William Smedley, in 1862, to compare traversing the waterways to Odysseus's battles with Charybdis and Scylla in Homer's *Odyssey*. Once the animals finally made it to the other side, if conditions permitted, overlanders turned the livestock out to rest, graze, and gather their strength before the group moved on to find a camping place for the night.[79]

Emigrants learned that listening to advice from people they met at the river should be considered carefully. In 1859 several Native Americans along the Green River warned an emigrant not to traverse the river where the man had intended to cross. As a result of the emigrant's unwillingness to listen, the cattle "went head over heels and in all kinds of ways, bellowing and making such a hideous noise in their struggles," causing some of the cattle to drown, some to return to the original shore, and only a few crossing to the desired side of the Green River. Joel Barnett, a witness to the disaster, concluded, "We felt sorry for the cattle but had little sympathy for the owner."[80] The lives of the animals, as well as those of the overlanders, depended on the emigrants learning to take advice and to work together to successfully maneuver across the large as well as the small rivers.

Mules, in particular, caused problems at river crossings. Considered "vexatious characters," mules often laid down in the middle of the streams, causing the provisions they carried to become thoroughly wet. Overlanders believed that the mules behaved in such a manner on purpose, and "if the wishes of our men could have been realized, mules would have been out of existence long ago."[81]

At several points on the Platte, Green, and smaller rivers, emigrants could take their wagons and livestock across the rivers on ferries that had been built by entrepreneurs. In most cases, the emigrants swam the livestock across the rivers but ferried their wagons. Frenchmen, Native Americans, Mormons, and others operated ferries on the rivers. Often of curious construction, ferries included boats, "a crazy mormon [*sic*] craft made of 5 canoes," and a ferry system with two pulley wheels fastened to ropes and attached to the boats.[82]

Between 1849 and 1859 the cost of ferrying the wagons and livestock across the rivers ranged between ten cents and seven dollars per wagon, averaging approximately four dollars on both the Platte and Green Rivers over the years.[83] For the owners of the ferries, significant amounts of money could be earned with the vast number of wagons, people, and livestock ferried each day. Overlanders had to wait several days to three weeks to cross at certain locations, such as the Loup Fork and Platte River junction and farther west along the Platte after Fort Laramie. At some points on the Platte, six ferries operated at the same time and location, serving five hundred wagons daily.[84] If ferry operators charged too much for the crossings, emigrants swam across the rivers with their wagons and livestock, avoiding the cost of ferriage. But basically the ferry owners could "charge whatever they wanted to and the emigrants had to pay it," particularly in areas where the depth of the water prevented the wagons and livestock from safely crossing. If the demand for the ferries in certain areas dropped, emigrants "struck a deal" with the owners for better fares.[85]

For those who either chose not to pay the ferry fees or could not afford to pay them, wagons could be rigged in such a manner to prevent them from sinking, and the overlanders then ferried themselves across the rivers. Those who chose to do the ferrying themselves needed to move away from the area where an entrepreneur had established a ferry landing. With the large number of people who needed to use the ferries, demands for government-run ferries rose that would, likely, charge less than the price of the independent operators.[86]

Just west of Fort Bridger, at the Bear River, tolls for crossing bridges ranged between seventy-five cents and one dollar per wagon. In one incident in 1859, Howard Williams encountered a man with a revolver demanding payment from Williams's group to cross the bridge. Williams paid the man but later found out that the bridge belonged to the US government and emigrants could use it for free. He hypothesized that the man on the bridge "ought to have been shot, and he will get it by some one [*sic*] yet."[87] Ferry operator rates west of Fort Bridger ranged between three and eight dollars per wagon, with the higher rates along the Snake. Even though considered "exorbitant" by the overlanders, the travelers paid the fees for the wagons but swam the stock as usual across the rivers. Like at other locations, the ferries came in a variety of construction styles, with one on the Bear River built of "roughly-built skiffs fastened together, side by side with board lain across, on which the wagons were placed," and at others ferry operators simply pulled wagons over the river using manpower. At some crossings on the Snake, ferry operators offered to swim the stock across the river for the overlanders at a cost of six dollars.[88]

After Fort Boise the overlanders began the four- to five-day ascent to the Burnt River, where they then slowly descended to the Powder River and found a good stream. Then at Flagstaff Hill they saw their first sight of the Blue Mountains. At this point the emigrants understood that their journey neared its end. At the Grande Ronde River they ascended long and steep mountains with crooked trails, causing them to continually shout to their teams "Gee" and "Haw" (right and left) in efforts to avoid the trees. Going up the Blue Mountains required six yokes of oxen per wagon, as the ascent proved "quite steep and precipitous." Once at the Umatilla River they found good grass but noticed that "the more sublime the scenery the more difficult became our progress."[89]

Only two hundred miles remained of the journey to Oregon City, the end of the Oregon Trail. If the overlanders had any of their cattle remaining by the time they arrived at The Dalles, they sent them along the pack trails where feed existed. The grasses along the narrow twenty-five-mile pack trail provided good feed for the animals and took one to three days to walk. Animals could also be wintered at The Dalles for a dollar and a half per head per month before continuing on to the overlanders' final stop.[90]

To avoid going over the Cascade Mountains, emigrants obtained passage on steamers that operated on the Columbia River. The boats also carried animals and wagons, but the expense of this service prohibited many from taking that option.[91] In 1846 Sam Barlow and a group of other early Oregonians built a ninety-mile road around the south side of Mount Hood that started in the Willamette Valley. He operated it as a toll road. He charged ten cents per head for each horse brought through the mountains. For those who chose not to use either the Barlow Road or the Columbia River options, overlanders took the trail through the Cascade Mountains. On the descent the emigrants chained the wagon wheels and slid the wagons down the trail or let the wagons down by ropes tied to a tree. The descent proved dangerous for all those concerned.[92]

While those going to Oregon confronted mountains with dangerous ascents and descents, those emigrants going to California faced a river that disappeared in western Nevada, a forty-mile desert, and trails through the Sierra Nevada. Two major exits from the Oregon Trail led overlanders to the California Trail. The first turnoff at Fort Bridger took the overlanders to Salt Lake City and then across the Hastings Cutoff to the Humboldt River in Nevada. In the second option emigrants followed the Oregon Trail past Fort Hall to the Raft River, where they turned southwest toward the Humboldt River, meeting the others who had turned southwest earlier.

The roads entering Salt Lake City from the north went up and down canyons and through narrow gorges where the animals "can scarcely keep a foothold." Several years later the roads improved, and emigrants found them "dry and good." Founding the city in July 1847, the Latter-day Saints built Salt Lake City at the south end of the Great Salt Lake. The Mormon community offered overlanders a place to rest and acquire new animals and supplies before continuing on to California. Mormons laid out the streets of the well-developed city in an east–west, north–south grid pattern, with houses built of unburned brick or clay that looked like "bee hives [*sic*]." Each home in the city had a garden, while the city's water came from a hot sulfur spring north of town.[93]

Many California-bound overlanders remained in Salt Lake City for a few days to refresh themselves and their livestock. Saints purchased worn-out animals from the emigrants and sold them fresh stock. Horses sold for $100 to $150 a head, or nearly double the cost of horses in the jumping-off towns. As elsewhere, the local entrepreneurs purchased the tired animals at a low cost, rested and fed the animals, and then sold them at a high price. No doubt, supply and demand dictated the cost of the animals for those needing them. Salt Lake City authorities paid citizens 25 cents for each loose animal they captured in an effort to prevent the livestock from eating the farmers' crops. The community fined overlanders whose animals got into the farmers' fields, including Finley McDiarmid, who paid $5 in "damages" when his horses ate in a local wheat field.[94]

Once rested and with fresh supplies, the overlanders left the City of the Saints and headed toward the Ruby Mountains in the Great Basin and then on to the Humboldt River in eastern Nevada. Having heard rumors about the dangers of the river and the desert that followed it, the preparations for the next five hundred miles to the Sierra Nevada could be likened to those at the start of the overland journey in Missouri or Iowa, only now the emigrants had experience but worn-out livestock.

Named by John C. Frémont after the naturalist Alexander von Humboldt, the river flowed approximately three hundred miles southwest into Humboldt Lake, an intermittently dry lake bed known as the Sink. At that point the land turned into a forty-mile desert. Overlanders found this part of the California Trail "the hardest part of the route for both man and beast" and the "meanest river of its length on earth."[95] In the desert, trails branched in two major directions, to the Carson River or to the Truckee River. If the emigrants took the Truckee route, they crossed into California at the Donner Pass, named after the emigrants who lost half of their group during the winter of 1846 while trapped in the mountains. If the emigrants took the Carson route, they crossed the mountains into Placerville, California. Approximately twenty-five thousand overlanders ventured on the Humboldt River trail, and with "good luck, or rather with good management," they hoped to cross the region with their goods and livestock intact.[96]

Following the Humboldt River called for much preparation during a time when the animals, as well as the overlanders, possessed little energy. At the start of the river in northeastern Nevada, the emigrants found that the Humboldt Meadows offered plentiful grasses, described as "superb" and "fine" throughout the years of emigration.[97] Emigrants likely exaggerated the quality of the grass because of the many recent miles of poor or no grasses available. Claiming the grasses to be "as tall as your waist" and the "best I ever saw," the emigrants heeded the warnings from earlier travelers and their guidebooks to gather grass, make hay, and collect water, as the conditions along the Humboldt deteriorated rapidly. While emigrants often had guidebooks and letters from friends or relatives to assist them in their planning for the cross-continent journey, others depended on the kindness of strangers they met along the way. William Cureton, in 1856, credited an "Angel" for telling his group about the dangers of the Humboldt River as they had never heard about the upcoming desert. The Angel told them that in order "to preserve your teams, you must fill all your surplus room with hay."[98] Cureton's group did as instructed and successfully crossed through the area.

While the overlanders gathered hay and collected water, their animals rested and ate as much grass and drank as much water as they wanted. As no good resting places existed between the Humboldt Meadows and the Truckee River or Carson River, the animals needed the time to gather their strength. Some overlanders lightened their wagon loads once again to make room for the water and hay for the livestock.[99] The overlanders needed the animals for the last few hundred miles of their journey and understood that the success of their endeavor continued to rest with the care they took of their livestock.

Much of the travel along the Humboldt needed to be during the overnight hours in order to avoid the sun's heat during the day. Guidebooks recommended that despite the urgency of crossing the desert as quickly as possible, the travelers needed to rest every two hours for ten to fifteen minutes throughout the night. At daylight the overlanders turned out the stock to permit them to gather as much dew as they could find to supplement the meager amount of water the overlanders provided them. After the early-morning break, they traveled until noon and then rested until late afternoon, when they started on the trail again.[100]

Between the Humboldt Sink, where the river disappeared into the earth, and the Carson River or Truckee River, the number of abandoned and dead animals grew steadily. Weakened by the lack of water, horses, oxen, and mules were abandoned, as the animals no longer possessed the strength to continue on and the emigrants had no water to give them. Overlanders expressed sadness when they abandoned their animals because they disliked "leaving an old friend" to fend for themselves in such an inhospitable region. In an effort to make the end of the animal's life as easy as possible, emigrants tried to find a bush to tie the animal to in a mistaken feeling that that might aid the animal in some manner.[101]

Overlanders assisted the abandoned animals they found as best they could. In order to entice the horses, oxen, and mules to follow the wagons, emigrants held out a few drops of water or blades of grass to the creatures, but in most cases the animals could no longer get up from the desert sands. Exhausted, starved, and parched, the animals died where they fell.[102] As the years of travel across the Humboldt increased, emigrants readily saw the bleached bones of animals on the roads from previous years and noticed that the livestock seemed to die in groups around abandoned wagons. Others found that the dead animals seldom decayed but "instead dried up like mummies" and "were surprisingly well preserved."[103] Overlanders found that if the dead animals "were laid side by side a person could walk across and not touch ground," and that they could not avoid seeing the remains of the livestock along the trail.[104] For those animals that had not yet decomposed, the stench that arose from their bodies furnished "no parallel" for this experience. Their "*nasal* knowledge" alerted the emigrants to the numbers of rotting animals they passed during the overnight hours and made their few hours of daytime sleep restless.[105] The animals that died between the Humboldt Sink and the Truckee River or Carson River numbered in the thousands. Few overlanders left the area without the loss of at least some of their livestock.[106]

"Poor grass, poor water, and hot days" well described the overlanders' time in the forty-mile desert. Even with the dangerous monotony of the Sink, another "great curiosity" revealed itself to the travelers. At Boiling Springs they witnessed water thrown to a height of three to four feet with much "hissing and rumbling." Some overlanders tried to make coffee with this water considered not fit for use, but never liked that drink again because of the memory of the flavor created by the minerals in the water. Before giving the hot water to the animals, it needed to cool to avoid burning the mouths and noses of the livestock.[107] For those who still owned cows, they discovered that the animals no longer produced milk and worried that the cows' teats "were squeezed so often they must have wondered what the men were trying to do."[108]

In the desert emigrants sold their extra supplies to lighten the wagon, while others sold their shares in the wagon company in order to unburden themselves of the responsibilities of caring for the animals and the goods they hauled. The sales allowed them to acquire a horse or a mule and become packers, choosing to walk the remaining miles while packing one or two animals with supplies. In 1849 John H. Benson sold his team and wagon to relieve himself of worry about water and grass for the oxen as well as the "anxiety I felt for the poor dumb patient animals."[109] Without the animals the travelers' lives improved, as they no longer had to concern themselves about the livestock, but the remaining miles would be more difficult, as they hauled their goods with only one or two animals or even on their own backs.

As teams came within a few miles of the river, the animals smelled the water and became unmanageable as they moved as quickly as they could, not stopping until they reached the middle of the river. The emigrants "rejoiced" for the animals and made no effort to restrain the horses, oxen, and mules, as they looked like they "would lie down to drink so eager were they."[110] Others temporarily abandoned their wagons and took their exhausted teams through ankle-deep sands to the rivers. Once refreshed, the overlanders and the teams returned to reclaim their goods.[111] In other cases, as the emigrants neared the Truckee River or Carson River, they found people selling water for one to two dollars per gallon, but the cost varied by the day. Selling water to emigrants became a commercial enterprise for those who had already crossed the desert and who understood the condition of the people and animals once they arrived on the other side of the Sink.[112]

The animals needed feed as well as water once they reached the rivers. Hay sold for 7.5 to 12.5 cents per pound, barley ranged from 25 to 40 cents per pound, and grass sold for 12.5 to 25 cents per bundle. The prices varied day to day, depending on availability and demand, as with everything the overlanders purchased along the miles. For those with the time and energy, cutting grass and bringing it to the edge of the desert became "the modus operandi" of earning money near the end of the trail.[113] By the early 1850s several trading posts existed in the Sink as well as on the Carson River. Emigrants suggested that the US Army should station soldiers along the Humboldt to service the needs of the emigrants, calling it a potentially "incalculable service to American citizens."[114]

After the hundreds of miles traveled thus far, the overlanders understood the danger to those who followed in their paths even a few days later.[115] The average emigrant spent two to four days in the desert between the Sink and the rivers. The dead animals the overlanders saw at the Sink admonished them to hasten along the trail, as they "are now getting a naked view of the *Elephant*" that breathed "out pain and misery in all ways and forms." In this region some overlanders began to "see the elephant," meaning that they had seen and experienced enough and sometimes even wished to return home, despite the miles they had already traveled. According to the overlanders, neither the deserts of Arabia nor the mind of Dante could have been as bad as the Humboldt River region.[116] They considered the area a "depot of carcasses" and as "fatal as the asp of Cleopatra's fruit basket." Once they passed through the area, they noted that they had finally crossed through Golgotha and Purgatory as well as over the Rubicon.[117] They had encountered nothing like the Humboldt Desert before, and coming near the end of the continental crossing, the journey proved especially difficult for humans and animals alike.

The journey to California did not end once they crossed the Carson River or Truckee River. The overlanders still had one to two hundred miles yet to travel to their destinations. The Sierra Nevada became the next obstacle that faced them. Nearly eighty miles wide at Lake Tahoe and possessing peaks eleven to fourteen thousand feet high, these mountains would have to be crossed by overlanders with little strength and by horses, mules, and oxen nearly worn out by the trials of the trails. A few of the emigrants sold their wagons and chose to ride one mule and pack another instead of facing the challenges of taking the wagons up and down the mountains, sometimes through snow six feet deep.[118] The majority took their wagons over the mountains using ten to twelve yokes of oxen to pull them over the slippery flat rocks. Six to eight men held the wagons upright with ropes, and many removed the wagon covers to lessen the impact of the winds in the mountains. If possible, they rested the oxen every two or three yards. Going one mile up the trail took nearly a whole day. Using a maneuver considered more dangerous than going up the mountains, overlanders used heavy chains or ropes fastened behind their wagons to lower the vehicles slowly down the trails.[119] Once they crossed over the mountains, wagon companies often broke into separate units, believing they no longer needed companionship and the safety of numbers. The lure of gold and farms called to the overlanders, and they did not desire to share their imagined wealth with others.

Since leaving the jumping-off towns, the emigrants had experienced and learned many things. They negotiated for ferry fees, found good camping sites, and ascended and descended mountains. Crossing rivers and learning whether to corral or picket the animals soon came easily to the overlanders. The landmarks they saw became milestones telling them how far they had come and how far they had yet to go. The daily tasks of the emigrants quickly became second nature to them. The importance of learning these things naturally meant that they could work with their animals more easily and efficiently, hopefully resulting in a successful cross-continental journey.

When the emigrants left their homes in the eastern United States or Europe, they clearly understood that the animals needed grass and water, but they did not anticipate the dangers in the water along the trail. They knew the basics of how to care for an animal, but did not realize that they needed to develop veterinary and blacksmith skills along the trails. More important, the overlanders never anticipated forming a relationship with the horses, oxen, and mules that accompanied them west. While climbing the mountains, fording the rivers, and crossing the deserts, the animals became a part, albeit it a brief one, of the overlanders' extended intimate circle. For the overlanders, the animals became more than motive power for the journey west.

Chapter Five

Caring for the Four-Footed Overlanders

On the journey from the Missouri River to the Pacific Coast, the overlanders developed a relationship with their four-footed companions that exceeded anything they anticipated and that surpassed their needs of using the animals to simply propel them westward. They did not foresee developing a friendship with the livestock that accompanied them on their journey.

The overlanders quickly learned the skills needed to care for their animals in order to keep them going as far and as long as possible. Shoeing horses and oxen, tending to sore backs and shoulders, and finding cures for alkali poisoning became part of the routine. Even if the overlanders used livestock in their home states and countries, the conditions on the trails required new skills. They felt compassion for the animals that needed to be abandoned along the way and mourned the deaths of the thousands of oxen, horses, and mules they passed by daily. From the beginning of the journey to the end, the demands placed on the emigrants to care for their animals never ceased.

Many emigrants possessed great faith in their livestock's ability to complete the journey, believing that the animals "seemed more like rational beings than dumb brutes" and that the livestock performed "every labor imposed upon them" with a desire to be of assistance to the overlanders.[1] Likely, they projected their feelings and apprehensions about the journey onto the animals. They understood that the livestock that pulled their goods would make their lives and adventures easier in the West and, as such, depended on them for their future.

Starting the journey feeling positive about the strength of their teams and that most of them "will go through like a top," initially many overlanders believed their animals to be better fit along the trails than in their home regions. They claimed that the horses and oxen "would barely subsist without working," but they quickly learned to favor the animals as much as they could.[2] The overlanders learned to walk alongside the wagons to lessen the weight pulled by the animals and to drive the animals slowly along the trail. As the horses, oxen, and mules weakened as the miles increased, favoring the animals became all the more important. When the grass and water deteriorated or became nonexistent, some emigrants even baked bread to substitute for feed for their horses.[3]

The physical deterioration of the animals began with poorly fitting saddles, harnesses, and yokes, as well as heavy packs. Most emigrants noticed that their animals became stiff as a result of the work expected of them and delighted when they could provide rest for the livestock. Rest proved to be more important to the animals as water and feed deteriorated, especially as they approached Fort Laramie and the Black Hills. The cattle had become worn out and weary by that time, but they had completed only one-third of their journey. With more animals becoming lame, many emigrants rested longer than during the first miles on the trails to allow the livestock to recover as much as possible before cracking the whips and moving on once more.[4] Many overlanders learned the delicate balance between giving the animals enough time to regain their strength and making sure that the groups moved forward at an appropriate speed to avoid problems in the western mountains as the season changed to autumn.

Despite the rest and care provided to the livestock, the overlanders noticed that the animals began stumbling along the trails. Soon after, they noticed that the animals lost weight and muscle and that their skin appeared "glued" to their bones.[5] Some emigrants felt a desire to help the animals regain their strength and believed that "their suffering was far greater than ours" along the trails. Claiming that their hearts "often ached" for the animals, some overlanders felt moved to tears as they watched the exhausted livestock march along the trails. Mary Rockwood Power, in 1856, wrote that "if anyone had spoken one word of sympathy [for the weary horses] I should have been a perfect baby" and cried for the animals' suffering. The overlanders wanted to help the animals but most often could offer only their "heartfelt sympathy" to them. Through the deserts or the mountains, it distressed some of the emigrants when they saw the livestock "always ready to do his best" despite being worn out. They often expressed disgust with other travelers who felt "no sympathy for the poor, patient animals" as they walked through the deep sands of the trails or suffered because of thirst.[6]

If the animals became worn out from the "fatigue of traveling" and could no longer pull the loads expected of them or even stand up, then the emigrants faced the dilemma of abandoning the livestock where they lay. Lameness and "feet worn out" caused many animals to be abandoned along the roads. Emigrants noticed the ravens that hovered overhead, "appearing to say 'we will have an opportunity to pick some of your bones before many days have gone by.'" Guidebooks recommended leaving the livestock where they fell because the emigrants' time "is of more value" than their livestock. The books also suggested that the Native Americans would take the abandoned animals if the animals recovered or eat them if they died.[7]

Throughout the journey and faced with little choice, the emigrants abandoned their livestock as necessary. Out of guilt at abandoning the animal or out of need for the animal's labor, some overlanders returned to the abandoned animals several days later, hoping to find them recovered and ready to return to the wagon trains. In 1859 Joel Barnett returned to where he had abandoned his ox Old Baldy. Luckily, he found the ox and took the animal to water and grass, and then they caught up with the rest of his group. Barnett later shod the ox to permit easier walking. He fared better than Margaret Frink, who, in 1850, returned for her abandoned horse only to find him dead. Harriet Scott Palmer lost a mare in the Cascade Mountains and luckily found it about one year later with a colt near Mount Hood.[8] Returning to the animals cost valuable time and energy as well as the labor of the animals that carried the overlander back to the abandoned one, adding to the miles the horse or mule had already traveled.

Occasionally, an abandoned animal found new owners. While the emigrants had no time to allow the animal to recruit, others behind them on the trails had the opportunity to pick up the abandoned, but now rested, animal. In 1852 Lodisa Frizzell spotted a head of stock with a note pinned to it stating that the cow had been left to die, but that if the animal recovered the former owner hoped that someone would take the cow but "not abuse her as she had been one of the best of cows" and deserved a good owner.[9]

The compassion expressed by some of the overlanders for worn-out animals deepened when they abandoned their livestock to an unknown fate. They understood that in many cases they had no other choice but to leave the animals behind because of the animals' physical condition. They felt that the horses, oxen, and mules "had faithfully served" the emigrants and that they could not force the animals to continue. Not all emigrants expressed or felt sadness at leaving their worn-out animals behind, but others admitted to shedding tears as they passed "those starved and worn out creatures that have rendered for the benefit of man their last staggering effort." The emigrants acknowledged the importance of the animals' efforts in their drive for new lives in the West and appreciated the haulers of their goods that "had even given us his very last step." An emigrant's cry of "another ox down" followed by the unyoking of the animal remained in the memory of many overlanders.[10]

After being unyoked or unhitched from the wagons, some of the worn-out animals attempted to follow the emigrants, but their efforts came to no avail, as they lacked the strength to catch up with the travelers. Once they collapsed on the road, other livestock gathered around their fallen companion and "gave utterance to a sad wail." The overlanders noticed that the abandoned animal "neighed and the other animals answered as if they understood" that the stronger animals must go forward without the worn-out ox, horse, or mule by their side. Many emigrants reported seeing the abandoned animals looking after them "with a look of resignation" as the wagon trains traveled forward.[11] No doubt putting human emotions into the experiences of the animals, the overlanders likely visualized their own experiences of leaving loved ones behind in their hometowns or burying companions along the trail.

To "terminate their sufferings" and prevent the livestock from having a slow death by starvation and lack of water, emigrants often shot the animals if they could. Others, believing that the animals might recuperate, allowed the animals to live because of a "mistaken feeling of humanity" that eventually caused suffering for the livestock as wolves killed the animals during the night.[12] Perhaps the overlanders hoped a group behind them might be able to help the abandoned animals, but they also believed that they had no alternative to the actions they chose.

As expected, the numbers of dead animals along the trails increased as the miles mounted. Along the Platte and Sweetwater, through the Black Hills and South Pass, and across the Humboldt Sink, a few emigrants kept tallies of the numbers of the dead "creatures" they passed every day. In May 1850 Finley McDiarmid began recording every dead animal he saw. He correctly claimed that his totals "will fall far short of the actual number." The dead animals lined the roads looking "as though resting," as the carcasses tended to dry up rather than decompose as in the eastern United States.[13] Using phrases such as "thick," "everywhere," and "scarce out of sight" to describe the innumerable dead animals, many emigrants found the stench emanating from the bodies "almost intolerable," but unavoidable and strong enough so "that a blind man might find his way by the odor of dead oxen." They also understood that the numbers of bodies they saw failed to indicate an accurate accounting of animals, as many of the livestock wandered away to die unseen by those counting them.[14] Few stopped to offer assistance to the dying livestock belonging to others, as each emigrant needed to take care of his or her own animals and could not afford to spend time, money, or feed on someone else's animals. Most emigrants only hoped that the animals could now "rest in peace."[15]

Even in death the livestock served the overlanders. Most obviously, the animals became meals, as fresh meat could not always be found. Nearing the end of the trails, especially along the Humboldt Sink, some emigrants carved out parts of their dead stock because their supplies of food had dwindled to nearly nothing. The meat became a source of income for some as the weeks stretched into months on the trail and the pocketbooks of the travelers emptied. In 1850, for example, ox meat sold for 12 to 37.5 cents per pound. Even the skulls of cattle and oxen became seats for the weary overlanders and served as message boards with directions and messages penciled onto them for those still behind. Many messages remained visible for several years.[16] In life and in death, the livestock served the needs of the emigrants.

In addition to abandonment and death, overlanders witnessed the abuse of livestock throughout the journey. Believing the beating of animals wholly unnecessary and cruel, guidebooks encouraged their readers to treat the animals kindly, just as did the domestic literature the emigrants read as children. US Army captain John Mullan wrote to "never maltreat them, but govern them as you would a woman, with kindness, affection, and caresses, and you will be repaid by their docility and easy management." Despite his warnings, at the jumping-off towns, the boatmen, likely hurrying to get the animals on board the ferries, sometimes injured oxen so badly that the emigrants needed to replace the animals during the first few days of travel. Overlanders often reported seeing teamsters whipping the cattle in their charge to the point where blood dripped down the sides of the animals. Calling his hired man a "Black Leg," Finley McDiarmid fired the teamster because the man whipped the group's horses unmercifully.[17]

As the overlanders had been taught by their ministers and the advice manuals, beating the animals caused the livestock to shrink from their approach as well as become "lazy and dull." Some emigrants threatened abusers with their revolvers, saying that if they whipped the animals again they would "blow him [the abuser] through." In 1853 Henry Allyn "ruminated over Paley's Moral Philosophy with great velocity" about the virtues of right and wrong when the man in charge of his group began to whip the mules. Not mentioning whether he "beat a little common sense into his empty skull," Allyn reported that the man no longer whipped the animals. Feeling sympathy for abused animals caused others to buy the wounded animals from the abusers and treat them kindly, realizing that if the beatings continued the animals would be dead before too long.[18]

Through good periods as well as bad, many overlanders attributed human characteristics and personalities to their horses, oxen, and mules. They credited the livestock with being good and sincere workers that tried their best and "never shirked" in their responsibilities in the jobs at hand. Some overlanders suggested that the animals offered their opinions of the demands made upon them by squinting and blinking at a task or refusing to do as the overlander demanded, sometimes resulting in overturned wagons. In areas with no water for many miles, emigrants tried their best to encourage the animals forward any way they could. To show their displeasure with having walked nearly thirty-five miles without water near Fort Hall, Daniel McLaughlin's mules looked at him "with a wink, as much as to say 'demnition [*sic*] tough, Mr. Ferguson'" before they started again. At other times the animals complained of the dust as if the owner had an actual conversation with them. Believing their livestock to have strong spirits, some emigrants praised the animals for being able to "keep up his end of the yoke" in a show of strength, ostensibly to please the overlander and demonstrate to other yokemates that the animal could still do the job at hand.[19]

Many overlanders enjoyed being with the animals and watching them play with their livestock companions when the conditions permitted. Sometimes the cows stole bread dough being prepared for meals even when plenty of grass existed for them to eat. The human observers claimed that the cows stole the dough to "torment an old lady" on purpose. They also noticed that male ponies enjoyed being around the mares and often neighed and sighed when out of sight of them, in much the same way a young man might pine when missing his loved one. In a similar vein, they believed that heifers could be lured across streams when the cows sighted oxen on the other bank, but that if the heifers changed their minds, they turned around in midstream, causing all sorts of confusion, including injuries to the animals.[20] No doubt because the emigrants missed their loved ones at home, they considered the relationships between the livestock a type of courtship and play.

Despite the friendly relationships that developed between the animals and the emigrants, rarely did the horses, mules, or oxen become the travelers' pets, creatures that they could indulge. Some pets, most often dogs, accompanied their owners west, and occasionally emigrants acquired animals along the way who served the purpose of pets. Cats, like dogs, came from home, while travelers adopted, for example, antelope and squirrels along the way. In 1849 James Philly's daughter refused to leave her home in Missouri unless her cat, Jip, joined the family, and reluctantly the father permitted the cat to come along. While the family traversed the Humboldt region, Jip brought a rabbit to the daughter's tent every morning to serve as her food, keeping her alive as she had been ill. In 1852 Eliza Ann McAuley acquired a squirrel and a young antelope as pets. Her antelope nearly caused a conflict between the Native Americans in the Bear River area and her family when the tribe's dogs killed her pet. The group offered skins and robes in exchange for the antelope, but the family refused the offers, explaining that they understood that the pet died as a result of an accident.[21]

Dogs dominated the pet population on the trails. Starting in the nineteenth century, dogs had become companions and members of the family. Emigrants often brought dogs with them, despite being uncounted in the tallies of western animal travelers.[22] Offering more to the overlanders than being simply objects of play and affection, dogs on the trails worked at a number of tasks. They herded livestock, keeping them moving "at a slow steady gait," and located missing cows and cattle that had strayed during the night. Perhaps more important, the animal served as part of the night watch, looking out for those interested in stealing horses or cattle. Dogs also protected the livestock from wolves and occasionally fought with the predators.[23] As helpful as dogs could be, they sometimes created trouble for the overlanders. A common complaint against them included their causing stampedes because of their barking and running among the cattle.

Emigrants sometimes shot at dogs for fun because they barked too much or because overlanders mistook the dogs for wolves or Indians.[24]

Walking along the trails, dogs suffered sore feet just like the horses, mules, and oxen. As a result, like their livestock counterparts, some dogs wore leather moccasins made specifically for them by their owners. If their feet continued to deteriorate, some owners put the dogs in the wagon for a few days until their feet improved. Dogs swam across rivers, occasionally receiving assistance from overlanders who tied ropes to the dogs' collars to prevent them from washing down the river or stream.[25] Emigrants treated the dogs as members of the traveling group and helped them as much as they could.

Like livestock, dogs became lost or ran off during storms, and some emigrants expended the energy of their horses to return to the last known location of the pets to see if they could be found. As with other missing animals, sometimes the emigrant succeeded in locating the lost dog. In 1851 Margaret White Chambers found her missing dog three weeks after a storm. When the dog noticed its family, Chambers explained, "what a rejoicing we had," and the dog went person to person, licking their hands. She claimed that "his delight was greater than ours." Her dog did not make it to the end of the journey, as he died crossing a desert, and Chambers lamented that if he had ridden in the wagon he "would have been all right," but they ignored him because her ill sister-in-law needed her attention.[26]

Calling their dogs "excellent" and "valuable," some overlanders praised the dogs as the animals continued to protect the owners even after the emigrants' deaths. Occasionally, overlanders reported seeing a dog lying on the owner's grave, and no inducement could lure the dog away from the site. The dogs even fought off wolves to prevent the wolves from digging up the bodies of their owners.[27]

Unlike their own dogs, emigrants found little worthy to be said about the dogs belonging to the Native Americans. Generally, they described the Indians' dogs as "wolfish" looking and possessing a howl that sounded "singularly mournful and plaintive." Amos Batchelder complained that the dogs do "not appear to understand the Yankee language, as we cannot make him take the least notice of anything." Despite that, they discovered that the Native American women and children "would not part with them at any price," being "much attached to them."[28] The overlanders noticed that as with their own dogs, Native Americans enjoyed the company of their pets as well.

Caring for the physical needs of the dogs and domestic animals involved handling situations often new to most of the emigrants. They needed to learn what could happen to the animals as well as how to cure any problems that might arise. Some of the overlanders lived and worked on farms in their home regions, and as such they likely knew a few ways to take care of the animals, but experience on the trails taught them how to care for the animals in situations unfamiliar to them at home. Much of what they understood came from the world of folk medicine and personal experience. Few professional veterinarians existed in the United States at the time of the overland movement. In 1850 the US Census reported forty-six veterinarians in the country, with twenty of them practicing in New York City, while others practiced in urban areas such as Philadelphia, Boston, and Chicago. Most likely, the majority of those men practiced farriery, or "the art of preventing, curing or mitigating the diseases of horses," and they also served as the "shoer of horses."[29]

Several publications about the care of animals existed in the United States at the beginning of the movement west, but whether the emigrants read them cannot be ascertained. Some of them included William Youatt's 1843 *The Horse,* published in Philadelphia, which provided a detailed explanation of diseases of the animals as well as a discussion of the medicines used in the treatment of them. Another publication potentially available to emigrants included the *American Agriculturalist,* which began publication in 1842 and advocated better care of livestock.[30] George Dadd, a pivotal figure in the development of American veterinary medicine, immigrated to the United States from Great Britain in the mid-1840s and in 1851 began publication of the *American Veterinary Journal,* which lasted only a few years. He sought information from anyone who worked with animals, not just veterinarians, because "there is much valuable information to be culled from all systems of practice." He desired to "fill up a blank long felt by owners of domestic animals" and to provide them with helpful information. Using his expertise he also published *The American Cattle Doctor* (1850), *The Modern Horse Doctor* (1854), and *Every Man His Own Cattle Doctor* (1856).[31] In 1850 Dadd matched the ideas of Enlightenment thinkers and Victorian-era writers as he advocated humane treatment for animals, stating that they "are as susceptible to pain and suffering as we are," and called for veterinary schools to end the anguish caused by scientific experiments.[32]

Often traveling before George Dadd published his works, the emigrants needed to make decisions regarding the health of their livestock with the best information available. Guidebooks for the journey west included information about whether the emigrants needed to shoe their animals. In his 1845 guidebook based on his 1842 journey, Lansford Hastings recommended not shoeing cattle because their "hoofs became more and more hardened" as they traveled, just as Joseph Ware's book recommended not shoeing oxen because the hot sand would "supersede the use of shoes." Regarding other livestock, Hastings recommended against shoeing mules, oxen, and cows, as that caused an unnecessary expense for the emigrants. Other guidebooks offered differing advice, such as J. M. Shively's, which recommended keeping the animals shod, and John Mullan's 1865 guide, which suggested shoeing only the forefeet.[33]

The emigrants discovered that despite the differing advice offered by the guidebooks, their experiences determined whether they needed to shoe their animals. Stone and gravelly roads often wore down the hooves of the animals, and as a result the livestock required shoeing in order to continue the journey. Possessing cloven feet, cattle and oxen needed two curved pieces of metal to shoe them. Some overlanders fashioned sheet iron into shoes for their cattle.[34] For those without the equipment or the expertise to shoe livestock, blacksmiths could be found at trading posts or forts along the trail or among the members of some of the wagon trains. Blacksmiths possessed some basic knowledge of animal health care as well, but the services most used by the overlanders involved shoeing the horses, mules, and oxen. In the 1850s the cost for shoeing horses or mules ranged between four and five dollars. Because of the cloven foot of the ox, the shoes for them ranged between six and eight dollars. Even the cost of shoe nails varied, between four and fifty cents per nail.[35] As with other supplies and services, the cost of shoes and shoeing varied with the availability of the goods and the negotiating abilities of the emigrants.

For those who either could not afford to shoe their livestock or believed shoeing unnecessary, the animals soon suffered from walking hundreds of miles on rough trails, on burned prairie grasses with dry sharp stubs, and through prickly pears with thorns. Many of them developed cracked hoofs that allowed grass and dirt to collect in the fissures. Many emigrants reported that the hooves started to break off the animals' legs. Called the "foot evil" by the overlanders, the injuries meant the hoofs retained "hardly any shape of a foot," and often death followed several days later. Most overlanders stressed the need to examine their animals' feet regularly and take immediate actions to help the livestock if possible.[36]

Compounding problems with the hoofs, alkali, saleratus, and selenium caused more problems for the overlanders and their animals. Little could be done about the alkali that endangered the hoofs of the animals, poisoned the water, and made the grasses dangerous to consume. Consisting of a soluble salt from the ashes of plants, alkali contained potassium or sodium carbonate. Large amounts of the substance made the soil infertile and the water undrinkable. The levels of alkali in the water depended on the availability of water, the soil chemistry, and the plant ecology in the region.[37] Guidebooks explained that the water in the middle of lakes or rivers contained less alkali because of currents. Many overlanders described alkali areas as having the color of strong ash lye, looking like lime; they also noted that the grass around an alkali area had a yellowish red color.[38]

Throughout the 1840s and 1850s alkali could be found along such rivers as the Platte, Sweetwater, and Humboldt. The guidebooks issued stern warnings about the regions and matter-of-factly stated that permitting the animals to drink the water there could cause death.[39] The large number of dead cattle and horses near an alkali source acted as a telltale sign of its presence as the overlanders passed through or camped near it. Preventing the animals from drinking the alkali-infused water proved to be hard work, as the livestock needed to drink at the end of a long day hauling the overlanders. During the overnight hours the emigrants often set up watches and guarded the animals as closely as they could to prevent them from drinking from the nearby rivers or ponds.[40]

Despite the precautions, animals ingested alkali and quickly became ill. For horses, at first the abdomen and the area between the forelegs swelled. Then the animal developed a cough that quickly destroyed the lungs, killing the horse. The swelling likely resulted from a reaction between alkaline material in water and gastric acid in the stomach. Horses, mules, oxen, cattle, and dogs all became victims of alkali consumption.[41]

In the same vicinity where the emigrants found alkali, they also found saleratus and selenium, both naturally occurring elements on the trails that caused troubles for the animals. Livestock became ill from saleratus, and the emigrants needed to move out of the areas as soon as possible to get to fresh water to allow the animals to drink as much as they wanted and then rest the teams.[42] Selenium, an essential nutrient for horses that helps with their immune system, can be toxic to animals and humans alike when consumed in high doses. Plants absorb the chemical, and the animals ingest it when they eat the plants. Because the plants are generally unpleasant tasting, the animals ate them only when they could find nothing else to eat, especially during the spring. Overlanders found that selenium caused the "blind staggers," where the horses wandered aimlessly and looked "like a drunken man." Emigrants noted dozens of dead livestock in the selenium areas.[43]

Not only did alkali, saleratus, and selenium cause problems for the wagon trains, but dust proved to be a serious problem as well, especially during periods of little or no rain. Complaints of the dust being "so heavy that it almost blinds us" lasted throughout the journey, whether along the Platte or Sweetwater River or in the mountains. Headwinds caused greater problems as the dust blew into the livestock's eyes. Emigrants reported trails covered with dust a "full leg deep," as well as half an inch of dust that coated the tops of the wagons, adding to the weight the teams must pull. In 1859 William Babcock and his fellow travelers cut open an ox that died suddenly. Fearing that they might find anthrax, they discovered that dust had covered the ox's windpipe and lungs. As a result of their finding, they decided to keep an extra few rods between their teams and bemoaned that they had not maintained greater distances between them earlier.[44]

In addition to dealing with the elements they encountered along the trails, emigrants needed to be able to identify and treat the diseases of livestock. Most commonly, they blamed murrain, an umbrella term for many diseases, including anthrax and foot-and-mouth disease, for the deaths of their animals. They cited hollow horn or horn-ail as a disease of cattle and oxen. Although hollow horn was not an actual disease, most emigrants did not understand that at the time, even though contemporary veterinarians published information about it. They believed that hollow horn caused the bony structure within the horn to collapse and shrink. To relieve the perceived problem, emigrants drilled into the horn in an attempt to cure it. Drilling into the horns posed the risk of penetrating the frontal sinuses, but believing it beneficial to the animals, some emigrants performed the procedure anyway.[45] Considering the emigrants' limited training in animal husbandry, they provided the best services that they could under the circumstances. They understood that the animals died from the effects of dust, a lack of safe food and water, and "by being worn out" by the tasks demanded of them. Some emigrants recommended that those following them should take along twice the number of oxen needed so that they could "favour weak cattle" and rest the teams as often as possible.[46] Keeping the animals healthy obviously aided the overlanders, but also showed their concern for the animals' welfare.

In the information available to the overlanders about the medical care of livestock, veterinarians recommended the use of medicinal opium for problems with horses, claiming that the narcotic acted as an antispasmodic, sedative, and astringent while also encouraging muscular strength. Despite the alleged positive features of the drug, they claimed that it worked best on Asian horses, which "seem to be sustained by its influence."[47] Doctors also advocated the use of the oil of turpentine as an "excellent antispasmodic." In 1854 Mary Burrell's group used turpentine and laudanum, a tincture of opium, to treat sixteen horses whose necks had become swollen from an undetermined cause. They applied turpentine to the outside of the necks and poured laudanum down the animals' throats. The swelling went down but "put the horses into torment." As a result, they hurried to a stream near the Carson River and for several hours poured water on the horses' heads and necks to remove the turpentine while slapping the animals' swollen parts with a paddle. Luckily, only two animals died, and they proceeded to cross the Carson River on their way to California.[48]

While using cures recommended to them, the overlanders also brought along home remedies to help the ailing animals. For snakebites Eliza Ann McAuley reported that her group successfully applied tobacco and whiskey to the bite on a steer's nose. Other snakebite cures included applying soap and lard to the wounds. For abrasions Captain Randolph Marcy advocated washing the bruises and sores with castile soap and water, but also noted that putting grease on the wounds or tying a bacon rind over the wound proved to work effectively as well. For colic the army captain suggested a mixture of brandy and laudanum poured down the throat of the animals, but if that failed, a mixture of water and dissolved chloride of lime could be substituted instead.[49] Basically, the overlanders tried whatever remedy they believed might help the animals.

Hoof and alkali problems brought forth a number of remedies. If not in an area with blacksmiths, occasionally emigrants tacked leather to the hoofs or tied the hides of dead livestock to the feet of the animals in need. Some emigrants brought tar or mutton tallow with them. First, they washed the foot with soap, scraped or cut off the "diseased flesh" from the hoof, and then poured boiling pitch or tar onto the foot. Applying it in the same fashion, emigrants heated mutton tallow and poured it on the hoofs. In 1852 Mary Jane Long noted that the animals "never flinched and the smoke would puff up off of their feet." She cried for the animals during these occasions, as she said "it was so hard on them to travel even when their feet were well."[50] Long, undoubtedly like many others, felt compassion for the animals and their demanding tasks on the trails and regretted being unable to do more for the livestock.

For alkali ingestion emigrants tried a number of remedies that needed to be administered as soon as possible to effect a successful cure. Bacon, acid, and vinegar taken alone or in combination proved to be the most common cure for alkali poisoning. Giving lard or fatty meats to the animal in need proved successful as well. Emigrants fried bacon to get the grease to give to their animals and sometimes threw the meat away, allowing others to claim the fried bacon for themselves.[51] Another version of the grease cure added vinegar to the fat or lard with a side benefit of curing diarrhea. Others put a large tablespoonful of tartaric acid at the back of the animal's tongue and then held its lips together to force the animal to swallow. Doses could be repeated if necessary. Eliza Ann McAuley, once again, recommended that a good dose of whiskey should be given to the animals that became "alkalized," although McAuley did not note why it worked. With the cures that included grease, the substance likely slowed down the gas-producing chemical reactions. Once the animals completed the treatments, the overlanders tried to rest the livestock, but noted that they remained "run down" for about two weeks after the event.[52] Despite the variety of cures used, the likelihood of them recovering completely seemed doubtful, given the high number of deaths in the alkali areas. Either that or not all overlanders knew the cures or did not possess the items needed for the remedies.

Many small health problems plagued the animals on the journey as well, and each difficulty cost the overlanders time and trouble. The problems often went beyond the ability of the emigrants to solve. They could not prevent the mules or cows from getting tangled in lariats or dying from bad water and poor feed. Dust and disease killed livestock every day, and the emigrants understood that. They wrote about being "very careful with them," spending many hours tending to the sick horses, oxen, or mules, and favoring "them all we could," but the animals still died. They expressed surprise at the cow that "was well and fat an hour before she died" and that "my finest oxen fell down and died in a few minutes." Some emigrants theorized that "some used little judgment and did not know how to take care of their teams" as reasons for the deaths. Others wrote poetry to explain the loss, saying that the "teams would weaken 'neath the loads they drew, sacrifice he must that surplus all."[53] Most overlanders tried their best to provide for the health of the livestock, but did not always understand why one cure worked and another did not. They needed the animals to pull their wagons and haul their goods, but they had also developed relationships with their animal traveling companions, making the loss of the livestock sadder than it would have been without that association.

After many miles and many trials, the overlanders arrived at their goal destinations in Oregon or California. Feeling thankful and a little surprised that they had made it, emigrants delighted that the animals "could be got through or at least the greater part of them." Some emigrants credited their skills as farmers accustomed to dealing with animals for the success that they had on the journey because "they provided for their horses before they provided for themselves," whereas they believed that doctors, lawyers, or "Town Pups" took care of their own needs before those of the livestock.[54]

While those going to Oregon often sold or traded off their animals at Fort Walla Walla or at The Dalles and took boats down the Columbia River to Portland or Fort Vancouver, California-bound overlanders often traded or sold their worn-out animals as soon as they arrived in Placerville (Hangtown). There, many emigrants traded their tired mules and horses for animals that the traders had purchased the previous season from emigrants coming from the Humboldt Desert. The traders purchased the run-down animals for $10 to $12 in provisions each and sold refreshed mules and horses for $150 to $200 each.[55] The emigrants had little choice, as they needed both new animals and goods.

Since the start of the journey, some overlanders knew that the animals that accompanied them would be worth "a big price" and be "money in your pocket" when they reached their destinations.[56] In California the prices received for the animals depended on the month the overlanders arrived. In August or early September, even horses and mules worn down by the trails sold for $50 to $250 each. Oxen sold for $150–$400 per yoke, while cows sold for $50 to $200 each. Whole outfits, including four horses, wagons, and harnesses, sold for $400.[57] Considering the cost of animals averaged $66 for mules and $70 for horses along the Missouri River, the sellers could at least break even or perhaps make a few dollars if they arrived early enough at Placerville. Later in the year, after more emigrants arrived in the community, the prices offered for worn-out animals dropped, as the market became glutted and the demand for recruited animals fell with the arrival of fewer overlanders. Despite that, the prices still ranged from $10 to $35 for worn-down, newly arrived animals. Placerville also offered pastures where cattle, horses, and mules could recruit and then be sold several months later for good prices, "as they pay high for flesh here, but nothing for bones."[58]

Many of the emigrants moved on to Sacramento and sold their animals at the Grand Horse Market on K Street. Daily, passersby heard "What do I hear for the Horse?" shouted out to get the attention of the buyers. Horse thieves also sold animals at the market that they had stolen earlier. Other communities in Northern California, such as Weaverville, had horse markets as well where emigrants could expect to receive $75 to $80, "a pretty good sale," for a worn-out horse.[59]

For those going to Oregon who took their animals on the ferries down the Columbia River or over the mountain passes in the Cascade Mountains, letters home and circulars told them to expect to obtain three to four times the original cost of the livestock in the sales at the animal markets. Although not receiving the quoted amounts, the prices nearly matched those in California, with a yoke of oxen selling for $150–$200, cows for $75–$125, and horses for up to $250. If unable to sell the animal because of its weakened condition, emigrants butchered it and sold the meat. In 1850 Isaac Harvey sold his steer's meat for 25 cents per pound, earning him $154, nearly quadrupling the $40 he paid for the animal in Missouri.[60]

With luck, the emigrants could wait until Portland or Placerville to sell or trade their mules, horses, and oxen. Prices offered for the tired animals often proved too good to turn down, even if the animals had become a companion on the journey. The time to end the friendship with the horses, oxen, and mules had come once they reached the West Coast. Now, the overlanders needed the animals for the price that they could get from their sale for the next part of their journey, that of mining, farming, or establishing a business.

The overlanders likely never anticipated that they would develop a personal relationship with their horses, mules, and oxen. Most of them understood that the relationship would be a temporary, practical one that lasted only until the destination or the death of the animal. Although sentimentality might suggest that they keep the animals as a reminder of their friends, family, and journey, practicality demanded that the overlanders bid adieu to their animals, as they needed the money from their sale. Just as the animals made the lives of the overlanders easier by hauling the goods, the animals' sale made the emigrants' new lives easier by providing for the next phase of their journey.

Why the emigrants cried at the loss or abandonment of their animals likely reflected a practical loss for the engines that pulled their goods and also the loss of their families and ways of life that they left behind in the East or in Europe. Whether the animals began the journey with them or not, they likely reminded the emigrants of the expectations that the overlanders had placed upon themselves when they undertook the adventure, that of striking it rich in the gold mines, on the farms, or in business enterprises. The animals brought the emigrants closer to a hoped-for success in the West.

While developing a companionship with the domestic animals, the overlanders discovered new animals at the same time. They saw animals that many had only heard about before leaving home, such as prairie dogs, buffalo, and wolves. While the domestic livestock brought emigrants closer to the riches of their new lives, wild animals symbolized a changing nation.

Chapter Six

Encounters with Wild Animals

Noticing, anticipating, and admiring the wild animals on the overland trails became as much a part of the emigrants' daily routine as picketing the livestock and finding good water and feed. The wild animals they encountered ranged from the smallest insects to the largest carnivores. Considering buffalo and antelope savory, wolves and mosquitoes scary, and prairie dogs saucy, most emigrants manifested a keen awareness and appreciation for the animal life that surrounded them on the Great Plains, through the deserts, and over the mountains. Many overlanders expressed sympathy for the beasts who sometimes died by the thousands at the hands of the emigrants. The mosquito caused many sleepless nights, while the buffalo encouraged many to try their hand at hunting for sport rather than for food. Prairie dogs elicited a happy delight from the overlanders. Even the sounds of the animals soothed or frightened the travelers. Some of the animals served them as sources of food and leather, while others symbolized their desires for new homes in the West.

The emigrants paid attention to wild animals throughout the journey. Aside from looking for meat for the evening meals, overlanders enjoyed seeing them, especially as many animals, such as deer and turkey, had disappeared from the eastern United States by the early 1700s. With the expansion of colonial settlers in New England, animal habitats dwindled due to the European ways of farming, and with the commercial value placed on animal pelts and furs the animal population correspondingly declined.[1] To help them know where to locate the wild animals, guidebooks by, for example, Joel Palmer, P. L. Platt and N. Slater, and Andrew Child told their readers wild turkeys could be found along the Little Blue River in Nebraska, buffalo between the Platte River and the Rocky Mountains, antelope throughout the Great Plains, and bear and elk in the valley of the Powder River.[2]

From the start of their journey, emigrants noticed the large numbers of ducks, geese, pelicans, squirrels, and rabbits along the shoreline of the Missouri River as they watched from the steamboats bound for St. Joseph. Familiar with many of Missouri's animals, they began to see creatures new to them once they crossed into Nebraska and followed along the Little Blue and Platte Rivers. In the Fort Kearny region, emigrants became "much diverted" by the wolves, foxes, antelope, and "prairie Rabbits" they encountered for the first time and had only "read about" before starting their journey.[3] Sometimes they could identify the animals right away, but in other instances the identity of the wild animal eluded them. Sometimes overlanders mistook large dogs for wolves and killed the dogs, thinking the animal might harm them. Animals new to the travelers also fell prey to the emigrants' weapons before they knew the animal's identity, such as in 1859, when George Averett and his fellow hunters rushed to see what animal they had killed. Some of the hunters thought the animal to be "one thing, and some another. Finally we decided it to be a wolf, a very vicious looking animal." Other unusual animals reported by the overlanders included a "small spotted Lemur" and an animal, likely a skunk, "as big as a grey hound [sic] with a wide, yellow strip at its back" that "stinked awfully bad."[4]

Overlanders enjoyed seeing animals new to them and watching the "dwellers of the prairie." They used adjectives such as *graceful, savage, shaggy,* and *harmless* to describe the wild animals they encountered.[5] Often daily, emigrants listed the types of animals they saw near the trails, such as elk, antelope, buffalo, prairie dogs, and birds, including sage hens, turkey, and plovers, and even fish and reptiles. Occasionally, they killed animals, such as prairie dogs, to "have a near view of them."[6]

Despite the many varieties of wildlife on the trails, overlanders most anticipated seeing buffalo for the first time. In Lansford Hastings's 1845 guidebook, he explained that buffalo would be sighted ten to fifteen days "from the States" and that "no adequate conception can be formed of the immensity of the numerous herds" that the overlanders would encounter. He told his readers that the plains and prairies would be "completely blackened with them, as far as the most acute vision extends." Despite his suggestions about the vast numbers of buffalo, not all emigrants immediately recognized the animal when first seeing it, as in the case of Sophia Goodridge Hardy who, in 1850, claimed, "We saw some animals, probably buffalo." On the other hand, J. D. Breyfogle, in 1849, recognized them right away and claimed that the men in his group "are in fine spirits thinking to get some fresh meat" after seeing buffalo for the first time.[7]

Commonly known as the *buffalo* and related to the European *wisent,* the American *bison* once ranged as far east as Virginia and North Carolina. Known since the sixteenth and seventeenth centuries to Spanish, French, and British explorers, the buffalo's range narrowed to west of the Mississippi River by the 1830s. Important to Native Americans on the Great Plains for the animals' meat, tallow, bones, and hide, few buffalo existed in the United States by the 1880s. In only fifty years, competition with Euro-American cattle and horses for food sources, an increase in bovine diseases such as anthrax and brucellosis, and the trade in hides created a rapid decrease in the buffalo population. Overlanders contributed to the deaths of many thousands of buffalo, but the commercial interest in the animal caused its demise. By 1894 the US Congress passed the first buffalo protection law, and a few ranchers, such as Charles Goodnight and Peter Depree, raised buffalo on their land. In honor of the once far-ranging bison, in 1913 the US Treasury coined the buffalo-head nickel, and in 1929 the Department of the Interior adopted the buffalo as its symbol.[8]

When the emigrants finally saw buffalo for the first time, many described them as an "imposing sight" whose "bellowings sounded like distant thunder."[9] They often compared the motion of the buffalo to the "undulating movement of a great sea as it rises in regular billows and falls in gently undulating troughs." The overlanders enjoyed seeing buffalo grazing with the antelope for many miles along the trail, as well as watching the "grace and stateliness" of the buffalo's swimming ability.[10] The numbers of buffalo that the emigrants saw astounded them. Many described the plains as "literally covered with buffalo" and said in some areas "no person could see through" the herds of buffalo as they moved across the land. Due to the large numbers of animals crossing before them, the emigrants had to stop on the trails and allow the buffalo to proceed on the animals' selected route.[11]

Emigrants needed to learn how to act in the presence of the animals which weighed between one and two thousand pounds each. Identifying a buffalo stampede and understanding the animal's posture when it slept became important tools to avoiding problems. In 1851 the British press reported that a buffalo tossed an overlander onto its back as he stood in the path of an oncoming stampede. Gaining control of the situation by dropping his legs astride the animal and grabbing onto the woolly hair on the buffalo's back, the Britisher rode it like a horse through a prairie-dog town and over three miles of plains before dismounting near an area with boulders, where he hid as the buffalo continued on its way.[12] As with learning the habits of the domestic livestock, emigrants also needed to understand those of the wild animals they encountered.

Cattle and oxen attempted to join herds of buffalo, and on occasion buffalo that lagged behind their respective herds tried to mix in with the overlanders' cattle. It took "a strict watch" to prevent their livestock from intermingling with the buffalo herds. Emigrants sometimes claimed that the domestic and wild animals shared a "sympathetic admiration" for one another, but that the buffalo mistrusted the cattle and oxen, while the domestic animals looked at the buffalo as "miserable cattle bosses," although the overlanders gave no reason for that conclusion.[13]

In addition to excitement, bison provided buffalo chips and hides for the use of the emigrants. Because of the scarcity of wood on the Great Plains, the dried excrement of the buffalo, commonly known as "buffalo chips" or "bois de vache," proved a "godsend" and a suitable substitute for wood for the campfires, as it burned like charcoal and made an excellent fire. Overlanders used buffalo chips, which was often gathered by children, for the first time as soon as they entered the plains.[14] Buffalo hides provided another way for the overlanders to keep warm. The robes could be purchased for about four dollars, and overlanders could sleep under or on top of them, providing a little comfort to the travelers. Buffalo robes also became funeral wraps for those buried on the plains. The skulls and shoulder blades of the bison provided a convenient place to leave a message for those following or a place to list the names and dates of those who had already passed. On occasion, the skulls served as a place to sit for the "weary traveler."[15]

As with buffalo, prairie dogs proved to be creatures the overlanders had only read about before the journey. Yet unlike buffalo which provided food, chips, and hides, prairie dogs possessed a lifestyle that most reminded the overlanders of themselves. Many emigrants described the animals in detail and believed the animals possessed civil governments, communities, and families. Not exactly sure where the prairie dog fitted into the animal kingdom, they considered the plains dwellers as "nothing more than an American species of marmots" or related to woodchucks.[16] Actually a member of the squirrel family, prairie dogs and dogs share no familial ties despite the names commonly associated with them.

Four species of prairie dogs exist in the United States, including the black-tailed, the white-tailed, the Utah, and the Gunnison. The fifth species, the Mexican, lives south of the Rio Grande. The most common prairie dog, the black-tailed, ranged from the forty-ninth parallel to the Rio Grande and from eastern Nebraska to Idaho in the northwest and to Arizona in the southwest. In the early sixteenth century Spanish explorers reported finding prairie dogs in Texas, while French explorers in the eighteenth century named them "*petit chien*," or little dog. By 1804 Meriwether Lewis and William Clark found them and called them "barking squirrels," switching to "prairie dogs" soon after finding them. Lewis and Clark sent a live black-tailed prairie dog to Washington, DC, where George Ord studied it and gave it the scientific name of *Cynomys ludovicianus*, the latter word Latin for Ludwig or Lewis, in honor of Meriwether Lewis.[17]

Overlanders discovered that black-tailed prairie dogs lived in the "driest and most barren land," weighed one to three pounds, and grew to fourteen to seventeen inches. Fascinated with prairie dogs, emigrants often provided long and detailed descriptions of the animals' physical attributes. Referring to the size of the animals, emigrants compared them to a "common ladies' lapdog" or a "half grown cat" and found them "chubby little fellows" with three-inch tails that moved "in a perpetual flutter" when excited. The prairie dogs' heads reminded the overlanders of bulldogs or young terrier puppies.[18] As the wagons and overlanders moved through the prairie-dog areas, the animals "kept up an incessant barking," with "short, quick, yet weak tones, like those of very young puppies," or like a "fiste [feisty] dog." Others called their bark a "peculiar chirp" and compared it to gray squirrels east of the Mississippi River. No matter the description of the bark, overlanders agreed that their sounds "filled the air" as the emigrants moved forward.[19]

Much "amused by the intelligence and acuteness" of the prairie dogs, some emigrants attributed to the animals "saucy" personalities, as well as a "sprightly mercurial nature." Lodisa Frizzell, in 1852, believed that if prairie dogs existed in Ireland, Irish legends of fairies and elves might be accounted for because the animals possessed "wonderful powers of instinct."[20] Because their interest in prairie dogs exceeded that of other animals, emigrants often wanted to capture "his dogship [*sic*]" to look at it up close. They also wanted to see the animal that moved so quickly it reportedly could dodge bullets from the emigrants' weapons.[21]

In 1806 Zebulon Pike discovered that prairie dogs lived in villages. During the 1840s and 1850s black-tailed prairie-dog towns or villages covered more than twenty-seven million acres, or forty-two thousand square miles of Kansas, Nebraska, South Dakota, and Wyoming. The emigrants reported much smaller ranges of one to five hundred acres, but, as they passed through only a small section of the community, they reported what they witnessed.[22] The trails passed through areas "literally filled" with prairie-dog holes or entrances to the subterranean cities or towns of the animals. Aboveground, the villages appeared to be laid out with "streets just as regular as a town of our own." At each entrance or mound emigrants reported seeing prairie-dog families "very conspicuously arranged as if for a free exhibition" and appearing to look "on Dress Parade."[23]

The well-maintained mounds leading into the prairie-dog tunnels looked like the mouthpieces of clarinets. While sitting at their entrances, the prairie dogs "chirped as birds might" at whomever passed through their area.[24] Upon entering the prairie-dog towns, emigrants sometimes theorized that the animals started barking either as a salute to the overlanders or as a warning to other prairie dogs of the overlanders' entrance into the community. When the emigrants camped in a prairie-dog town, the barking of the animals prevented many travelers from getting a good night's sleep.[25]

Despite the appearance of a community made up solely of prairie dogs, other animals such as rattlesnakes, owls, deer mice, and cottontail rabbits lived in the dens and tunnels as well. Emigrants did not understand the relationship between the members of the underground community, but they knew about the danger of thrusting their hands into the tunnels in search of prairie dogs that had been shot. Although the prairie dogs were "not fastidious in their choice of companions," as Francis Parkman discovered in 1846, William Cureton, ten years later, praised their acceptance of others living in their community, believing that "they never had any family rows." Cureton's views likely reflected his rocky relationship with his own family in Missouri and with his cousin, who twice asked the fourteen-year-old to leave the wagon train halfway to California.[26]

Just as Cureton attributed human family qualities to prairie dogs, other overlanders credited the plains dwellers with the ability to form governments. Perhaps because the prairie dogs lived in large communities and offered no threat to the emigrants, the overlanders attached values and desires to the prairie dogs that many of them likely hoped for themselves in the West. Emigrants described wild and domestic animals in detail, but no other animals fascinated them as much as the prairie dog, making their ideas about the small creatures unique in the overlanders' ideas about animals.

Emigrants noticed that the prairie dogs ran from hole to hole and appeared to be discussing "the affairs of the world." They claimed that prairie dogs regulated "the body politic," possessed "systems of civil government and domestic economy," and called for the use of sentinels to alert the community with their barks of approaching dangers.[27] Finding aspects of the prairie-dog society republican in nature with laws designed for the "public good," those emigrants who observed the prairie dogs also believed them to be "purely Democratic in which the females have an equal voice." Some overlanders considered the prairie dogs "restless and on the move," implying an expansionist tendency among the little animals, especially as the emigrants understood that when the prairie-dog communities grew too large, they sent out "colonies sufficiently strong to form villages," perhaps reinforcing their own notions of Manifest Destiny.[28] Considering the publication of John L. O'Sullivan's 1845 article that advocated American expansion, the idea that overlanders saw their own community reflected in the society of prairie dogs should not come as a surprise.[29] With the War of 1812 a recent memory and the Mexican-American War ongoing during the early days of the push to the Pacific by the overlanders, many emigrants likely saw a similarity between wagon trains that invaded the prairie-dog towns and the emigrants' experiences with the British and Mexican wars with the United States. As a result, they appeared to understand the prairie dogs' desire to "vindicate the invaded rights and insulted dignity of the republic" by protecting their homelands and wanting to add territory to their domains.[30]

While some overlanders considered prairie dogs akin to themselves in wanting a well-run society, they found wolves and coyotes dangerous and little more than "the gleaners of deserted camps," living off the scraps emigrants left behind. Wolves weighed 50 to 130 pounds and grew to four to six feet in length and three feet high at the shoulder, while coyotes, sometimes called prairie wolves, grew to sizes only slightly smaller than their cousins. Emigrants considered wolves "sly and cowardly chaps" and reported seeing them nearly every day along the Platte River, across the Humboldt Desert, and through the mountains. Some believed that wolves considered humans a large animal and as such would not attack them, while others found that wolves "seemed not to fear us" and would attack people when the wolves found a person alone. That belief led Lucia Darling, in 1863, to fear for the safety of her sister, Pauline, who had wandered away from the encampment. Calling her sister's name as she searched for her, Lucia heard only "the howling of wolves which was not very pleasant." With the help of others from her group, Lucia found Pauline safe before too much time had passed.[31]

The fear of wolves caused emigrants to raise an alarm whenever they sighted the predators. In 1846, just east of Fort Laramie, Heinrich Lienhard found six wolves near him. He attempted to scare them away by suddenly jumping up from a crouched position, but instead of being chased off, the wolves "beheld my capers with stoic calm." Lienhard credited the "nobility" of the wolves rather than his antics for the animals' decision not to attack him. Others hoped for the wolves to come near enough to their campsites to "make them a leaden present," but the animals generally failed to come within range of the rifles.[32]

In some places on the Great Plains, hunters had to compete with the many wolves to get to a downed buffalo first. Overlanders attributed the large number of wolves in the desert and in the mountains to the growing numbers of abandoned livestock as the journey neared its end. After consuming the dead animals, the wolves and coyotes left the skins behind, allowing a few emigrants to take advantage of the hides to make sandals.[33]

Regarding those who died along the trails, emigrants hoped that wolves would not disturb the burial sites, but acknowledged that "a thrill of horror" arose in those who buried their dead. To prevent the wolves from uncovering a corpse, most overlanders tried to bury the body as deeply as possible and to place large rocks and boulders upon the grave.[34] Despite those efforts, wolves managed to dig up the graves and fight for "choice seats at the feast of human flesh." Emigrants found partially consumed human arms, tendons, and legs strewn across an area in "a sad spectacle." Dr. Thomas Flint, in 1853, picked up human bones that he found but chose not to report his findings, "for fear some of the boys might feel superstitious about it."[35]

The howling and gnashing of the wolves' and coyotes' teeth during the nights caused a "dismal noise" and "a chilly sensation and almost cause [*sic*] the hair to raise on one's head." The scavengers' "midnight revelry" provided a "free concert," but the cracking of their teeth "made night hideous." From sundown to sunup, the wolves howled near the camps of the overlanders, causing some to be "hushed to sleep" by their calls, while others slept little for fear of the animals. Concern for the livestock rose as the area became a "howling wilderness," likely reminding some emigrants of what their ministers had taught them about the days after the collapse of the Garden of Eden. The wolves and coyotes readily ate the dead oxen and buffalo they found near the camps of the emigrants.[36]

Dogs, as well as wolves and coyotes, belong to the genus *Canis*. While emigrants considered wolves and coyotes enemies, the domestic dogs that traveled with the overlanders served as friends as well as workers. At the start of the journey, dogs frequently barked at and chased the wild animals they encountered. Emigrants reported that prairie dogs often wore out the domestic dogs as the prairie dogs dashed about their villages, avoiding capture. More practically, domestic dogs served as hunters, even slowing down a buffalo so the dog's owner could kill it. If buffalo needed to be moved out of the way of the oncoming wagons, dogs chased the much larger animals away, with the buffalo fighting the dogs to maintain their place. Dogs also served as guards for the overlanders, warding off hungry wolves and coyotes during the nights. But like their wild relatives, the dogs barked while the wolves and coyotes howled, much to the displeasure of their owners.[37]

While overlanders learned to deal with animals unfamiliar to them, such as buffalo and prairie dogs, they also encountered other new animals, including antelope and grizzly bears, that bore some resemblance to animals from the eastern United States. Emigrants often spoke excitedly about seeing antelope for the first time, noting that they looked like deer from their homes, and theorized that the animals could be the "connecting link between the common deer and the goat." To reassure himself about the identity of the creature, Sarah J. Cummins's husband, in 1845, consulted a natural history book and decided that the animals resembled descriptions of antelope. The emigrants enjoyed watching the antelope "gliding over the plains" and seeing them "sailing in the air" and described the animal as the size of a deer and as "fleet as a greyhound," resulting in its being difficult to shoot.[38]

As with antelope resembling deer in the East, grizzly bears reminded overlanders of black bears. Grizzly bears ranged from the Great Plains through the mountains, weighed eight hundred pounds, and stood five to eight feet tall. A formidable enemy if encountered accidentally, bear were sighted throughout the journey. The bear had a footprint of nine by thirteen inches, and a bear's roar echoed with a "beastly thunder." Many overlanders reported being less than two hundred feet from a grizzly but choosing not to shoot it. Not all the travelers understood the dangers involved when encountering this bear, such as John Rueger from Switzerland, who noticed a lot of red berries on a bush near the Feather River. He first saw the berries but quickly noticed eleven-inch bear tracks. Failing to recognize the potential danger in the situation, he continued to gather berries without worry. Rueger reminisced that at that time, he knew nothing of the grizzly's might, or he "would not have felt so secure."[39]

While fearing bears, wolves, and coyotes, mosquitoes, "this little contemptible insect," caused the emigrants much grief and many sleepless nights.[40] Overlanders found mosquitoes along most of the journey from eastern Nebraska through Idaho. With mosquitoes swarming in large clouds along such rivers as the Platte and Bear, emigrants reported that they could hardly see the setting sun because of the number of insects. Emigrants complained that the mosquitoes prevented them from "appreciating Nature's grandeur" after they set up camp, except for the "music" created by the insects' buzzing.[41] The insects "feasted" at the expense of the overlanders and looked "like they've been fed well." Calling the mosquitoes their "genuine friends" because they "stick to us," many overlanders claimed the insects to be the same size as "Italian bees" or as large as "Eliphants [*sic*]." The mosquitoes fell upon their prey like "famished tigers," with bills "never blunted by use."[42]

After setting up camp for the night, emigrants had trouble guarding their livestock because of the constant attacks by the insects. With the mosquitoes seemingly intending to cause problems for the travelers, the overlanders christened rest stops "Camp Misery."[43] Even with good grass and roads nearby, mosquitoes forced the travelers to "hitch up and toddle" during the middle of the night to find better, or at least mosquito-free, campsites. While some emigrants traveled late into the night to find another camp, others left early in the morning, traveling until the sun heated the atmosphere sufficiently to get rid of the insects long enough to have a quiet breakfast. Undoubtedly, mosquitoes transmitted malaria to some emigrants on the trails, yet few understood the insects' impact on their health and may not have associated the bites with the resulting illness later.[44]

Mosquitoes attacked domestic animals just as often as they did humans, worrying the emigrants who wanted to protect their livestock as much as they could. The discomfort inflicted by the multitudes of mosquitoes prevented the horses, oxen, and mules from grazing, even though the livestock had not eaten in many hours. Completely covering the domestic animals, the mosquitoes caused the horses and mules to lie down, kick, and roll on the ground in an effort to rid themselves of the insects. Emigrants often noted that the animals felt the attacks of the mosquitoes as much as the humans did.[45]

Overlanders devised a number of techniques to lessen the impact of mosquitoes on themselves as well as on their livestock. They tried to grease the horses to keep the "bloodsuckers" from attacking the animals, and they created a camphor mix that seemed to work. They used smudges to produce smoke that would hopefully banish the insects, but often that had little impact on the mosquitoes. If overlanders traveled in small groups, they tried to camp near a larger group because the mosquitoes appeared to gravitate toward the greater supply of blood, leaving the smaller group at least a little safer from attacks. With some luck, a wind might arise that carried the mosquitoes away, or safety could be sought at higher elevations.[46]

Along the same trails as mosquitoes, other insects regularly tormented the emigrants and their livestock. Wood ticks attached themselves "with such tenacity" that pulling off the body left the head under the skin, creating painful wounds. "Out of all proportion to their size," a sand fly's bite resulted in "pimples" covering the emigrants' faces. The sand fly also found its way under wristbands, behind ears, and under hats, leaving bites in its wake. "Roaring like a distant wind," beetles descended upon the overlanders in clouds "so numerous as to almost overpower us" and thumped against the sides of the tents during the nights. Just past Fort Bridger many emigrants reported that grasshoppers and shield-backed katydids, mistakenly identified as crickets and called "Mormon crickets," covered the ground. They noted the size of the "crickets" as three inches long and three-fourths of an inch in diameter. Calling the insects "cannibals," the tiny creatures covered the road for miles, and the livestock, wagons, and overlanders crushed them at every step. Near Fort Kearny horseflies proved troublesome, as their bites to the livestock made the animals' blood run. The only insect that seemed to please the emigrants proved to be the firefly, which looked like "flying fire sparks" during the evenings.[47]

While insects flew or crawled, reptiles and amphibians slithered or hopped along the trails. Overlanders found snakes of many species, including rattlesnakes, common along the entire route, but especially on the Great Plains and in the Humboldt Desert. Early on the journey nearly all of the emigrants decided that "no mercy to these creatures can be shown" because of the danger the snakes posed. Travelers found snakes concealed along the bluffs and even in the campsites. Occasionally, snakes surprised the overlanders as the emigrants drank from the Platte by swimming beneath the thirsty emigrants, causing the drinkers to have a "speedy rising" to avoid the creatures. Bathers ran screaming from the water "in their scanty bathing attire" when they spotted snakes swimming nearby. Overlanders could identify rattlesnakes and garter snakes, but generally described others as "striped snakes" or "hissing vipers." Emigrants used little hesitation in killing the reptiles and occasionally noted that they "had quite a fight" with the snakes before shooting them with their pistols. Upon entering the Humboldt Desert, emigrants sometimes left impaled snakes near the trail to serve as warnings to the oncoming travelers of the number of snakes in the area.[48] As with snakes, overlanders found lizards along the trails, sometimes in areas where no other creatures lived. Often confusing the horned lizard for a horned toad, the emigrants found the lizard to be a "curious living creeping hopping thing" that seemed to have an attitude that said "I'm ready for you." Others thought it looked like a "link between the toad and the terapin [*sic*], or mud turtle," but all agreed on its swiftness.[49]

The overlanders also noticed the creatures that flew overhead. The "feathered animals" included, for example, geese, ducks, cranes, plovers, and eagles.

The overlanders noticed that many of the birds from their home regions did not exist in the West, such as robins and bobolinks, but that larks, blackbirds, ravens, thrushes, and cuckoos could be found in abundance. They enjoyed hearing the larks singing merrily and the "Curlee! Curlee! Curlee!" of the curlew, as well as watching the prairie grouse "strutting with high heads" as the birds walked along. On the Great Plains they noticed the light chirping of distant birds during the evening or morning.[50]

While seeing new creatures brought joy or fear to the overlanders, they hunted wild animals for pleasure and for food. Emigrants often hunted prairie dogs for "curiosity & amusement," as the animals proved a challenge to hit because they popped their heads out of their tunnels and quickly withdrew them. Citing amusement in their efforts to kill prairie dogs, overlanders sometimes spent hours attempting to hit them as they popped in and out of their holes, and in the process wasted a lot of ammunition that could not be replaced as they went west.[51] The desire to signalize themselves as good hunters encouraged some to kill wild animals, such as buffalo or wolves. The skins or hides of the animals served as evidence of the kill but added weight to the wagons that the livestock pulled.[52]

As the hunters sought out their prizes, they noticed a creature that reminded them of the rabbits from home but had longer ears and tails. In the same scientific family as rabbits, hares, also known as jackrabbits, looked like fawns to the overlanders and were more beautiful than the rabbits from home. Emigrants noticed the animals across the Great Plains and into the Blue Mountains.[53] The overlanders also spotted beavers as they crossed the South Pass and moved deeper into the mountains. They admired the dams of the beavers, finding that the animals displayed much ingenuity in building their homes.[54]

While overlanders considered hunting buffalo as "the greatest sport that can be conceived of," they understood that killing the animal provided a "fine mess of beef" for the emigrants.[55] Guidebooks explained that "fresh meat can be procured at any time" and that "here is meat, kill and eat." Although enthusiastic in their assertions about the availability of game on the trails, emigrants needed to be skilled in killing the wild animals that they sought.[56] During the first weeks along the trail, emigrants hoped to develop the skills needed to kill the buffalo.

While exhibiting a great deal of excitement about hunting buffalo, the emigrants claimed that the animal possessed an "aspect the most diabolical" and proved difficult to hunt because of its unpredictability and its becoming a "perfect picture of mingled rage and terror" during the chase. Emigrants reported being chased or even attacked by the buffalo. Often using twelve to twenty bullets to kill the bison, some overlanders reported that the bullets seemed to rebound off the heads of the animals as if they had hit a rock and that the shaggy coats of the buffalo seemed "impervious" to the rifle shots.[57] While buffalo provided excitement, hundreds of pounds of meat, and a desirable hide, emigrants often failed to get their game. Buffalo swam across rivers, such as the Platte, to escape their pursuers, or the dead animal fell into a river and the overlanders could not get it out. At other times nearby wolves took possession of the newly killed animal.[58]

Emigrants tried deception to lure other game within range of their rifles. They tied red handkerchiefs to their shotguns to raise the curiosity of antelope to bring the animals close enough to kill, but shooting at antelope often proved as futile as shooting at buffalo because they outran the rifles aimed at them. Bear also proved elusive, as the "'varmints' were used to the tricks of travelers and vamoosed over the hills" and avoided the overlanders' shots. In 1852 John Hawkins Clark, a bear hunter from western Virginia, promised himself that if he failed to get the next bear he saw, he would "break his rifle over a wagon wheel."[59] Whether Clark got his bear remains unknown.

Despite the successful hunts for wild animals, some of the guidebooks warned overlanders not to waste time hunting game, as it could endanger the long-term goal of getting to their destination. Emigrants quickly learned that trying to overtake game endangered the health of the ofttimes poorly rested and fed horses that were needed to carry them to the Pacific. Domestic animals sometimes suffered injuries on the hunts as well, such as being shot accidentally or falling during the chase.[60]

In the emigrants' efforts to make sure that they secured meat, they wounded or killed more animals than they required for their needs. Occasionally, the wounded animals wandered off and went unfound by the hunters. In other cases, hunters succeeded in shooting several animals and, being unable to carry back all the meat, took only the choicest parts or even left whole animals behind for the wolves.[61]

Some overlanders acknowledged the "great waste" that occurred during the hunts, but believed that little could be done to alleviate the excessive number of animals killed. In an attempt to lower the number of wasted kills by members of the Church of Jesus Christ of Latter-day Saints, Brigham Young, president of the church, issued a commandment that decreed that the Mormons may kill only the number of animals that they required for their use. Despite his announcement, some church members, like non-Mormon men, wanted to signalize themselves by killing a buffalo. Their fellow travelers suggested that "they had better learn wisdom" and kill only the animals needed for their sustenance. As more Latter-day Saints continued to join their brethren in Salt Lake City, they strengthened Young's rules about killing more animals than needed. They may have been unaware of the commandment issued by Young, but they understood the significance of the slaying of so many wild animals.[62]

Primarily, emigrants killed wild animals to provide meals for themselves, not to demonstrate their prowess in slaughtering game. Overlanders found a "plentiful supply" of buffalo, antelope, bear, rabbits, prairie hens, and prairie dogs. Commonly, they described wild game as a "treat." Attributing their fondness for game to its being fresh and not like their salted provisions, they concluded that the meat "was decidedly relishable."[63] When they shot a buffalo, they feasted while "luxuriating upon" the meal before them. Many found the meat "very fine and sweet," while others considered buffalo similar in quality to beef. Despite the meat's favor among overlanders, guidebooks and ministers warned them not to consume too much because overeating might cause diarrhea.[64]

In addition to eating the buffalo right away, emigrants often jerked the meat. With some slight variations, the jerking process called for the meat to be thinly sliced and laid on a scaffold above a slow fire to dry and to smoke it. That process took about a half day. Others dipped the meat into brine and then hung the strips over a frame, allowing the moisture to drain. If the emigrants could not spend the time at a campsite, they tied string to the meat strips and hung it inside the wagon near its top. By the end of a day's travel, the dry climate had jerked the meat. Processing the meat in such ways resulted in the buffalo meat lasting for several months.[65]

Prairie dogs also found their way onto the emigrants' menus. They discovered that the burrowing rodent "furnishes a delicate meat" that some found "quite palatable" and as "good to eat as chicken." Not all favored the prairie-dog meat, so they "ate sparingly," while others "could not get them down."[66] Responses to the prairie-dog meat may have varied from other wild animals because the emigrants had anthropomorphized the creatures.

Antelope, bear, and hare proved to be other treats for the emigrants' meals. While some compared antelope to venison, others found it "really toothsome" and the "best meat" they had ever eaten. Emigrants enjoyed the "delicious odor of frying steaks" and believed the "fragrant flesh" of the antelope "was quite a luxury for us."[67] As was the case with antelope, many emigrants ate bear for the first time on their way west. Needing multiple shots to kill the bear, overlanders found that "Bruin steak was not at all a bad article" for their meals.[68] Like buffalo, antelope and hare proved to be a desirable change from salted meats and provided an excellent meal.[69]

In addition to the wild meats, fowl and fish supplemented the overlanders' meals. Sage hens, "much finer looking" than prairie hens and about the size of a small turkey, could be found on the Great Plains and into the Sierra Nevada. Opinions about eating sage hen varied more than with other meats. For example, in 1851 Lucena Parsons enjoyed sage hen because the birds tasted of sage, while Amelia Hadley held a diametrically opposite view of the bird's meat, precisely because the mountain sage "scents their flesh." Hadley concluded that a skunk would be preferable to the sage hen for her meals. In addition to eating the bird for meals, some overlanders ate the sage hens because they believed it might help cure sore throats.[70]

Overlanders found plentiful fish along the trails. Finding trout in such rivers and streams as Spring Creek, the Big and Little Sandy Rivers, Weber's Creek, Green River, and Bear River, emigrants considered the fish the "finest I ever saw" and the "first speckled trout we have seen." For some, the trout proved to be "very shy" and did not bite, leaving them without a fish supper for the evening.[71] In addition to trout, the emigrants found a good-tasting whitefish near Heber Spring that looked like alewives. Salmon could be found the farther west the overlanders went, especially in modern-day Idaho and eastern Oregon. Using fish hooks or rifles, emigrants occasionally caught salmon that weighed thirty pounds each, while others caught enough salmon to load two mules with the fish in just over an afternoon of fishing.[72] Native Americans along the Snake River dried and packed salmon and readily traded it for manufactured goods, such as clothing, beads, calico, and knives. They also willingly exchanged salmon for bread and sugar.[73]

Hunting buffalo, fearing wolves, and enjoying prairie dogs resulted in many of the emigrants developing an awareness of the impact they had on the wildlife along the trails. As the years went on, they noticed that the buffalo and antelope had grown wary of the approaching wagon trains and had moved away from the trails as the number of overlanders, wagons, and livestock continued to mount. By the early 1850s emigrants readily found that game had become scarce along the trails, perhaps because of "the many rifles that are aimed at them," as well as the expanding regions of poor-quality feed for the wildlife to eat.[74]

Not only had the wild animals grown cautious of the overlanders, but some emigrants came to realize the "sad havoc" they had perpetrated on the native animal populations. They noticed the wanton destruction of the animals as the number of bones along the Platte River grew throughout the 1840s and 1850s. They claimed that the "ground was fairly white with the bleached bones" of the dead animals.[75] Some emigrants regretted the slaughter of the "creatures that God has made," claiming that it "seems wicked" to kill an animal just to prove one's ability to do it. The killing of wild animals, even for food, elicited emotional responses from some of them. When in 1842 an overlander shot the head off a prairie dog, he noted that "there was a feeling in the little incident, a something human, which raised the animal in my estimation, and ever after I did not attempt to kill one of them, except when driven by hunger." Even the chopping down of a tree to see if there was honey in a beehive caused regret for Heinrich Lienhard, when he discovered that neither bees nor honey existed in the tree, but instead only two black snakes lived there. Regarding his action, he concluded that "we had robbed those harmless creatures, wholly uncalled for, of their safe home."[76] Although the emigrants needed to eat wild animals to survive, killing the creatures both large and small caused many of them to reflect on the damage done to the animal populations.

With the death of so much wildlife, emigrants blamed the "march of civilization across the continent" for the fewer number of animals available for meals and wondered how future overlanders would survive the journey. In her 1905 reminiscence Sarah Watson Hamilton wrote a poem that addressed the missing game:

> The forward trains no doubt found game ahead,
>
> But the trains behind they did not spare,
>
> But drove the timid creatures from the road
>
> And it was seldom that we got a share.[77]

Emigrants also understood that the destruction of the wildlife hurt the food supplies of the Native Americans as well. They suggested that their dead cattle be left behind for the "half starved red man" because the overlanders killed the game that existed in the area, leaving little for the local population.[78]

While wondering about what those behind them might eat, many of them came to appreciate many of the wild creatures they met on the journey for more than their meat or hides. In their letters and diaries, they often wrote romantically about the animals they encountered along the Platte River, describing, for example, buffalo as "awe-inspiring" and the delight they had in "seeing them in a state of nature." The memories and feelings about the animals remained with them for many years after their journeys, such as Joel Barnett, who nearly seventy years later reminisced about the "noisy little chipmunks," the "chirping bluejays," and an owl "wiggling his head."[79] The emigrants regularly commented about the "beautiful and playful antelope," the "prowling wolf," and the music of the prairie dog's bark.[80]

Overlanders traveled hundreds of miles to their destinations. They had been told by friends, relatives, and guidebooks to expect to see, eat, and fear wild animals along the way. They found the new creatures fascinating for their behavior, their sounds, and the resources they provided. The overlanders learned to hunt, to bury their dead well, and to simply enjoy some of the creatures they encountered. The wildlife ranged from the two-thousand-pound buffalo to the lightweight mosquito. They sympathized with the animals whose range and herds became smaller and smaller as the settlers moved west. The emigrants learned that the animals provided more benefits than simply providing meat for their stew pots and frying pans. During the long journey many overlanders developed an appreciation of and compassion for the wild animals they encountered on a daily basis, finding some savory, some scary, and some saucy.

Chapter Seven

End of the Trail

By the time the overlanders reached the end of the trail in California, Oregon, or Utah, many of them discovered that they had paid more attention to domestic and wild animals than they anticipated when they left home. They understood from the start that they would hunt buffalo and antelope for their meals and that they would need to watch over their livestock to make sure they remained healthy. Some of them expected the deaths of many horses, mules, and oxen, but they never envisioned developing warm feelings for the animals that shared the journey with them.

Crossing the continent took a toll on both the animals and the emigrants. Many of them lost their lives in the quest for land, religious fervor, or gold. The labor of pulling the wagons and carrying goods and people proved to be too much for some of the livestock. Many overlanders also realized that the thousands of people, wagons, and domestic animals negatively impacted the habitat of wild animals, including their favorites, buffalo and prairie dogs. The buffalo population declined rapidly during the years of the overland migration, and by 1890 only eight hundred animals remained in scattered herds on the Great Plains. In their turn-of-the-century reminiscences, B. F. Nichols bemoaned that "our Government was so slow in protecting those noble animals and preventing their extermination," while A. A. Enos found it difficult to believe that the buffalo, which "roamed these prairies in such vast numbers, should now be so nearly extinct." Enos explained the loss, citing "the law of civilization and advancement" and calling it "the survival of the fittest."[1] Prairie dogs elicited an emotional response at the demise of their population as the overlanders marched through the animals' territory. In 1852 Lodisa Frizzell proposed that prairie dogs barked "some very hard things against us, for the boys shot several of them" for sport rather than for food. She wrote that "it is pitiful to see them when one is wounded or killed outside, & cannot get into his hole; others will rush out, & drag him in . . . as if they had been informed, & understood that one of their number was wounded or dead."[2] Some emigrants also commented that the wolf population increased in the Humboldt Sink area because of the large numbers of abandoned or dead livestock that lined the trails as a result of the rush for new lives.[3]

Most emigrants worried about themselves first, not the travelers who followed them. Occasionally, the overlanders acknowledged that those who came after them would have a hard time finding good grass and water for their livestock because each group took what that group needed, not considering or worrying about the needs of those behind them. With each advancing year and with the tens of thousands heading west, the amount of available grass deteriorated, forcing many overlanders to take their livestock several miles from the campsites in order to find good food and water. The extra miles the animals traveled to obtain nourishment added a considerable burden to the livestock's already taxing job of pulling the wagons on the day's march. Without a doubt, the emigrants who followed months and years later faced greater misery than those who traveled earlier.

When they began the journey, the emigrants carried new ideas about animals with them to the West, ideas they had read in the domestic literature, heard in the sermons, and even saw in the art they viewed. For many of them, the notions of treating the animals kindly became practical guidance and something that needed to be done on a daily basis. Few of the nineteenth century's advocates of kindness considered wild animals in their works, but despite that the overlanders easily adapted what they had learned to the new situations they encountered. The emotional attachments to animals, wild and domestic, demonstrated by many of the emigrants likely resulted from the experiences they shared with the animals, such as deprivation of water and food or the dependence upon one another for survival to the end of the trail.

While most overlanders successfully made it to their destinations using animals and wagons, others became "foot packers," or kept only one or two mules. Those who made this decision abandoned the wagons because the "mules kicked, horses ran off, [and] ropes stretched," causing many daily difficulties that took too much time away from the day's travels.[4] As the journey proved more arduous than many expected, some emigrants encouraged others still in their homes to forgo the overland journey, with its troubles and inherent animal problems, and take the sea route instead, as they "will rue" their decision to travel overland to their destination.[5] Some even spoke of "seeing the elephant" and a desire to go home, as the experience of going west proved too much for them. Ironically, they used the name of an animal to describe their feelings about the traveling experience.

The animals became "as essential" to the overlanders as "the ship to the mariner," and as with ships, animals needed to be maintained and cared for along the journey.[6] In order to do that, the overlanders provided, to the best of their abilities, water and food for their animals. Many rested on Sundays with a new appreciation for the Bible's words demanding that the people rest their animals, as well as themselves, on the Sabbath. Some emigrants also came to believe that their horses, oxen, and mules felt pain at the loss of the livestock's animal companions. The overlanders cared for their animals as best they could, even under extremely harsh conditions, such as those in the Humboldt's forty-mile desert.

Some of the emigrants grew to "love the patient and hardy" animals they worked with for many months. They developed a companionship with their animals, such as John Hawkins Clark, who claimed, in 1852, that the ears of the mules "do not seem half so long as they did at the commencement" of the journey, while A. J. McCall would not sell Charley, his horse, as he did not consider himself a "mercenary slave-trader" and that the "wealth of the world could not have tempted me to part with one drop of this precious blood." Even William Smedley, who had not yet reached the end of the trail, refused to trade a lame ox for a sound one because his fondness for his animal companion had become so strong that he declined the offer.[7]

For most of the overlanders, their relationship with domestic animals lasted only three or four months if the animals survived to the end of the trail. Experiences with wild animals lasted for only as long as the overlanders remained within the range of the animals' home territories, such as prairie dogs or antelope on the Great Plains. By the end of the trail, some overlanders sold their worn-out animals at the nearest animal markets, while for others the animals stayed with the emigrants for many more years. Sometimes the animals who traveled west with the overlanders accompanied the emigrants on a return journey to their eastern homes, such as in 1866, when Joel Barnett and his family returned to Iowa from Oregon after having spent seven years in the West. As before, the Barnett family traveled by prairie schooner pulled by horses and mules, including Old Charley, the horse that had made the initial journey to Oregon.[8]

Although not scientists, many emigrants came to enjoy the animals they traveled with and those that they encountered. They did not study the animals; instead, they lived with them and depended on them for the animals' abilities to haul the goods and the families and to provide meat for their meals. Some emigrants also came to understand that the thousands of domestic livestock, wagons, and people might have appeared to the wild animals as "the invasion of a formidable enemy," seeking to conquer the new lands of the West.[9]

Most of the overlanders made it to their destinations, while the livestock and wild animals suffered heavy losses. The overlanders came to a new understanding and appreciation of the domestic and wild animals they worked with and encountered. They learned that success did indeed depend on their relationship with their animals and that the teachings of many of the ministers, philosophers, and Victorian-era writers may have influenced how they treated the animals on the trails. Without a positive relationship between the emigrants and the animals, a successful journey likely would not have occurred. In addition to pulling the heavy loads west, the livestock became the emigrants' traveling companions, helping the travelers create new lives in the West. Like the emigrants, the domestic animals became overlanders.

Notes

BL Bancroft Library, University of California–Berkeley
CHS California Historical Society, San Francisco
HL Huntington Library, San Marino, CA
MMRL Merrill J. Mattes Research Library, Independence, MO
OHS Oregon Historical Society, Portland
USHS Utah State Historical Society, Salt Lake City
UU University of Utah, Special Collections, Salt Lake City

Preface

1. See, for example, John D. Unruh Jr., *The Plains Across: The Overland Emigrants and the Trans-Mississippi West, 1840–60;* and David Dary, *The Oregon Trail: An American Saga.*

2. Alfred W. Crosby Jr., *The Columbian Exchange: Biological and Cultural Consequences of 1492;* Virginia DeJohn Anderson, *Creatures of Empire: How Domestic Animals Transformed Early America,* 4. Other excellent books focusing on animals stress the importance of the creatures as machines or pets. These include Ann Norton Greene, *Horses at Work: Harnessing Power in Industrial America;* Clay McShane and Joel A. Tarr, *The Horse in the City: Living Machines in the Nineteenth Century;* and Katherine Grier, *Pets in America: A History.*

1. Going West

1. Sara Alexander, "A Little Story of the Experiences of Sara Alexander When Crossing the Plains in 1859," MS A22, USHS, 1.

2. Joel Barnett, *A Long Trip in a Prairie Schooner,* 10; "Importance of Veterinary Science."

3. Delford Benton, "Diary," May–June 26, 1850, MS 85/4C, BL; Overton Johnson and William H. Winter, *Route Across the Rocky Mountains,* 213.

4. Richard White, *"It's Your Misfortune and None of My Own": A New History of the American West,* 189–90; Robert V. Hine and John Mack Faragher, *The American West: A New Interpretive History,* 185–86; Derek Hayes, *Historical Atlas of the American West with Original Maps,* 92–93.

5. Reginald Horsman, *Race and Manifest Destiny: The Origins of American Racial Anglo-Saxonism,* 219; Hine and Faragher, *American West,* 199.

6. Absolom B. Harden to Brothers and Sisters, 1848, MS 11, #007972, OHS; John S. Barker, "Journal of an Overland Route to California," March 16, 1850, MS OCTA, Box 2 Ba–Bl, File Barker, MMRL.

7. Ray Allen Billington, *The Far Western Frontier, 1830–1860,* 70, 154; Hine and Faragher, *American West,* 182–86.

8. Lansford W. Hastings, *The Emigrants' Guide to Oregon and California,* 6; G. W. Thissell, *Crossing the Plains in '49,* 13.

9. Hine and Faragher, *American West,* 202; Dary, *Oregon Trail,* 244.

10. Ray Allen Billington and Martin Ridge, *Westward Expansion: A History of the American Frontier,* 229; Hine and Faragher, *American West,* 234.

11. Dary, *Oregon Trail,* 72–74; Arthur King Peters, *Seven Trails West,* 91; Billington, *Far Western Frontier,* 92–93.

12. Dary, *Oregon Trail,* 85–87.

13. Unruh, *Plains Across,* table 1, 119; table 2, 120.

14. Hine and Faragher, *American West,* 193.

15. See, for example, Capt. John Mullan, *Miners and Travelers' Guide to Oregon, Washington, Idaho, Montana, Wyoming, and Colorado via the Missouri and Columbia Rivers;* Capt. Randolph B. Marcy, *The Prairie Traveler: A Hand-Book for Overland Expeditions;* and Francis Parkman, *The Oregon Trail.*

2. Emigrants Take More than Livestock West

1. Paul Waldau, "Animals," 356, 358.

2. Ibid., 360; James Rachels, "Animals and Ethics," 274; Sophia Menache, "Animals in the Bible and Qur'an," 77; Norm Phelps, *The Longest Struggle: Animal Advocacy from Pythagoras to PETA,* 55; V. Anderson, *Creatures of Empire,* 52.

3. Menache, "Animals in the Bible and Qur'an," 76, 77; Waldau, "Animals," 361; V. Anderson, *Creatures of Empire,* 52; Rachels, "Animals and Ethics," 274.

4. Mary Sayre Haverstock, *An American Bestiary,* 19–20; Menache, "Animals in the Bible and Qur'an, 77.

5. Waldau, "Animals," 358; V. Anderson, *Creatures of Empire,* 52; Katharine M. Rogers, *First Friend: A History of Dogs and Humans,* 88.

6. Katherine C. Grier, "'Kindness to All Around': The Changing Ethics of Animal Treatment in the Middle-Class Household, 1820–1870," 182; Haverstock, *An American Bestiary,* 23; Charlotte E. B. Tonna, *Kindness to Animals; or, The Sin of Cruelty Exposed and Rebuked.*

7. James Turner, *Reckoning with the Beast: Animals, Pain, and Humanity in the Victorian Mind,* 45; Laurence Pringle, *The Animal Rights Controversy,* 4.

8. John Bigland, *A Natural History of Animals,* 20, 187.

9. Rev. J. H. Avery, *The Land of Ophir, Ideal and Real: A Discourse Delivered at Austinburg, Ohio, Before a Company About Proceeding to California,* 3–4.

10. Gerald E. Jones, "Zoophily in Mormonism," 7; Gerald E. Jones and Scott S. Smith, *Animals and the Gospel: A History of Latter-day Saint Doctrine on the Animal Kingdom,* 7, 8, as quoted in Gerald E. Jones, "The Gospel and Animals."

11. Jones and Smith, *Animals and the Gospel,* 7.

12. Gary L. Francione and Anna E. Charlton, "Animal Advocacy in the 21st Century: The Abolition of the Property Status of Nonhumans," 17; V. Anderson, *Creatures of Empire,* 53.

13. Phelps, *Longest Struggle,* 26, 27, 34.

14. Ibid., 34, 35; James Serpell, *In the Company of Animals: A Study of Human-Animal Relationships,* 123; Rachels, "Animals and Ethics," 274.

15. McShane and Tarr, *Horse in the City,* 8; Linda Kalof, *Looking at Animals in Human History,* 97–98; Rachels, "Animals and Ethics," 275; Pringle, *The Animal Rights Controversy,* 5; Lilly-Marlene Russow, "Changing Perceptions of Animals: A Philosophical View," 34; Serpell, *In the Company of Animals,* 136.

16. Rod Preece, *Brute Souls, Happy Beasts, and Evolution: The Historical Status of Animals,* 3–4; Pringle, *The Animal Rights Controversy,* 6; Francione and Charlton, "Animal Advocacy in the 21st Century," 17; Diane L. Beers, *For the Prevention of Cruelty: The History and Legacy of Animal Rights Activism in the United States,* 21.

17. Benjamin Franklin, *Poor Richard's Almanack,* 22; Thomas Paine, *The Writings of Thomas Paine: The Age of Reason, Part I.*

18. Grier, *Pets in America,* 131; Katherine C. Grier, "'The Eden of Home': Changing Understandings of Cruelty and Kindness to Animals in Middle-Class American Households, 1820–1900," 317–18.

19. Daniel Walker Howe, "American Victorianism as a Culture," 527; Russow, "Changing Perceptions of Animals," 33; Grier, "'Eden of Home,'" 319.

20. Arthur M. Schlesinger, *Learning How to Behave: A Historical Study of American Etiquette Books,* 17–18; Grier, *Pets in America,* 132; Joan Goodwin, "Lydia Maria Child"; Poetry Foundation, "Lydia Huntley Sigourney."

21. Grier, "'Eden of Home,'" 317.

22. Grier, "'Kindness to All Around,'" 182, 184; Grier, "'Eden of Home,'" 317, 322; Lydia H. Sigourney, *Letters to Mothers,* 98; Turner, *Reckoning with the Beast,* 45.

23. V. Anderson, *Creatures of Empire,* 8; Tonna, *Kindness to Animals;* Turner, *Reckoning with the Beast,* 9; Humphry Primatt, *A Dissertation on the Duty of Mercy and Sin of Cruelty to Brute Animals,* 322.

24. Grier, "'Eden of Home,'" 316, 322, 329; Aaron Honori Katcher and Alan M. Beck, "Animal Companions: More Companion than Animal," 265; Grier, *Pets in America,* 12.

25. L. Child, *The Mother's Book;* Tonna, *Kindness to Animals;* Sigourney, *Letters to Mothers,* 35; Bigland, *Natural History of Animals,* 20.

26. Mrs. L. G. Abell, *Woman in Her Various Relations: Containing Practical Rules for American Females,* 58; Tonna, *Kindness to Animals;* Beers, *For the Prevention of Cruelty,* 24.

27. Sigourney, *Letters to Mothers,* 98; Tonna, *Kindness to Animals;* Catherine E. Beecher, *A Treatise on Domestic Economy,* 351–52.

28. Milton W. Brown, *American Art to 1900: Painting, Sculpture, Architecture,* 177; H. Barbara Weinberg and Carrie Rebora Barratt, "American Scenes of Everyday Life, 1840–1910"; David S. Reynolds, *Waking Giant: America in the Age of Jackson,* 278; Daniel J. Boorstin, *The Americans: The National Experience,* 239.

29. Barbara Groseclose, *Nineteenth-Century American Art,* 10–12; David M. Lubin, "Art in an Age of National Expansion: Genre and Landscape Painting," 94; Haverstock, *An American Bestiary,* 147; Mary Sayre Haverstock, "An American Bestiary," 60.

30. Haverstock, *An American Bestiary,* 19, 23, 25, 33; Haverstock, "An American Bestiary," 40; V. Anderson, *Creatures of Empire,* 45, 53, 64; Jon T. Coleman, *Vicious: Wolves and Men in America,* 37, 41.

31. Haverstock, "An American Bestiary," 40, 44; Haverstock, *An American Bestiary,* 13.

32. Clara Endicott Sears, *Highlights Among the Hudson River Artists,* 177–78; Haverstock, "An American Bestiary," 60; Haverstock, *An American Bestiary,* 147; Michael David Zellman, comp., *American Art Analog,* 160.

33. Marshall B. Davidson, *The American Heritage History of the Artists' America,* 62; Boorstin, *Americans,* 237.

34. Edward F. Dolan Jr., *Animal Rights,* 16; Beers, *For the Prevention of Cruelty,* 20–21.

35. Pringle, *The Animal Rights Controversy,* 7; Beers, *For the Prevention of Cruelty,* 23; Dolan, *Animal Rights,* 16–17.

3. Preparation for the Journey West

1. Harmon Mothershead, "River Town Rivalry for the Overland Trade," 14, 15, 22.

2. Unruh, *Plains Across,* 68–70, 73; C. W. Smith, *Journal of a Trip to California, Across the Continent from Weston, Mo., to Weber Creek, Cal., in the Summer of 1850,* 16; Finley McDiarmid, letters to his wife, Constantia McDiarmid, 1850–51, May 23, 1850, MS C-B 605, BL.

3. John Mack Faragher, *Women and Men on the Overland Trail,* 25; William E. Parrish, Charles T. Jones Jr., and Lawrence O. Christensen, *Missouri: The Heart of the Nation,* 156; Lodisa Frizzell, *Across the Plains to California in 1852,* 8; C. Smith, *Journal of a Trip to California,* April 8, 1850, 13; Howard D. Williams, "A Diary Kept by H. D. Williams Containing a Record of the Events of a Journey Across the Plains," MS A2746, USHS.

4. Faragher, *Women and Men on the Overland Trail,* 24; Lavinia Honeyman Porter, "By Ox Team to California," 198, 201.

5. Billington and Ridge, *Westward Expansion,* 104; Parrish, *Missouri,* 71–72; Eugene T. Wells, "The Growth of Independence, Missouri, 1827–1850," 35.

6. Billington, *Far Western Frontier,* 99.

7. Unruh, *Plains Across,* 68; Joel Palmer, *Journal of Travels over the Rocky Mountains,* 144; Dary, *Oregon Trail,* 112; Robert J. Willoughby, *Robidoux's Town: A Nineteenth-Century History of St. Joseph, Missouri,* 26–27.

8. Unruh, *Plains Across,* 115; Frank S. Popplewell, "St. Joseph, Missouri, as a Center of the Cattle Trade," 444; Amasa Morgan, "Diary," April 16, 1849, MS 2001/111, BL; Dr. J. S. Shepherd, "Journal of Travel Across the Plains to California, and Guide to the Future Emigrant," 1850, microfilm F593.S45, BL, 2.

9. W. J. Lusk to editor, *Frontier Guardian* (Kanesville, IA), May 9, 1850; Unruh, *Plains Across,* 72; Faragher, *Women and Men on the Overland Trail,* 25.

10. Mothershead, "River Town Rivalry," 14; Parkman, *The Oregon Trail,* 15–16; Phoebe Goodell Judson, *A Pioneer's Search for an Ideal Home,* 13.

11. Charles Albion Tuttle to Maria Tuttle, correspondence with his wife and miscellaneous papers, 1849–88, April 6, 1849, MS C-B 427, BL; Margaret A. Frink, "Adventures of a Party of Gold-Seekers," April 23, 1850, 75; Morgan, "Diary"; George Belshaw, "Diary," May 15, 1853, MS 1508, #007892, OHS.

12. George McKinley Murrell to Mary Ann Murrell, correspondence, 1849–54, MS HM 36338–36403, HL; F. McDiarmid to C. McDiarmid, May 23, 1850; Barker, "Journal of an Overland Route to California," March 16, 1850 (see chap. 1, n. 6); William R. Brown, *An Authentic Wagon Train Journal of 1853: From Indiana to California,* 18; Amos Batchelder, "Journal of a Tour Across the Continent of North America from Boston via Independence, Missouri, the Rocky Mountains to San Francisco, by the Author in 1849," May 3, 1849, Accn. 1085, UU; Maria Parsons Belshaw, *Crossing the Plains to Oregon in 1853,* May 14, 1853, 15.

13. Alonzo Delano to *Free Trader* (Ottawa, IL), 1849, in his *California Correspondence;* Chester Ingersoll to editor, "Overland to California in 1847: Letters Written to the Editor of the *Joliet [IL] Signal,*" MS HM 66492, HL; Dr. J. Tarbell, *The Emigrant's Guide to California,* 3; Heinrich Lienhard, "California Immediately Before and After the Discovery of Gold," MS 89, Book 1, UU; Shepherd, "Journal of Travel," 1.

14. Hastings, *Emigrants' Guide,* 145; Philip L. Platt and N. Slater, *Travelers' Guide Across the Plains upon the Overland Route to California,* xvii; George W. Buchanan, "Letter and Circular of Information for Prospective Emigrants to Oregon," 307 (letter to William M. King, March 2, 1847); Johnson and Winter, *Route Across the Rocky Mountains,* 213; Lienhard, "California Immediately Before and After the Discovery of Gold," 3; Marcy, *Prairie Traveler,* 26; Murrell to Sally B. Murrell, April 8, 1849.

15. Batchelder, "Journal of a Tour Across the Continent"; H. Williams, "Diary," April 25, 1859; Platt and Slater, *Travelers' Guide Across the Plains,* xvii; Frink, "Adventures of a Party of Gold-Seekers," April 23, 1850, 75; Harry Bradley and Nancy Jane Bradley, "Diary," May 3, 1852, MS OCTA, Box 3, File: Bradley, MMRL; W. Brown, *Authentic Wagon Train Journal,* May 17, 1853, 18.

16. Unruh, *Plains Across,* 406–8.

17. Ingersoll to editor, "Overland to California," May 14, 1847; Batchelder, "Journal of a Tour Across the Continent," May 5, 1849, 5; Marcy, *Prairie Traveler,* 41; Horace Barton Pomroy, "Account Book," April 6, 1850, MS C-F 128, BL; Joseph E. Ware, *The Emigrants' Guide to California,* 5; Hastings, *Emigrants' Guide,* 146.

18. J. M. Shively, *Route and Distances to Oregon and California,* 4; J. Palmer, *Journal of Travels over the Rocky Mountains,* May 7, 1845, 12; Tarbell, *Emigrant's Guide to California,* 18; Andrew Child, *Overland Route to California,* vii.

19. Hosea B. Horn, *Horn's Overland Guide to California*, 73–74; Buchanan, "Letter and Circular of Information," 310 (to editor of *Western Expositor*, February 15, 1847); "Horse and Ox Shoeing," *Frontier Guardian* (Kanesville, IA), April 17, 1850; Parkman, *The Oregon Trail*, 16; Henry Atkinson Stine to Emily Stine, letters and journals of Henry A. Stine on his overland trip to California from St. Louis to Sacramento, May 8, 1850, MS Little, Box Sp–Sy, MMRL; Shepherd, Journal of Travel *Across the Plains to California*, April 19, 1850, 6.

20. Batchelder, "Journal of a Tour Across the Continent"; O. Allen, *Allen's Guide Book and Map to the Gold Fields of Kansas and Nebraska and Great Salt Lake City*, 6; James Mason Hutchings, "Diaries," June 1, 1849, MS 69/80C, BL; John Hawkins Clark, *Overland to the Gold Fields of California in 1852*, May 6, 1852, 232; Buchanan, "Letter and Circular of Information" (to editor, *Western Expositor*, February 15, 1847).

21. Thissell, *Crossing the Plains in '49*, 9. See also, for example, Buchanan, "Letter and Circular of Information" (to editor, *Western Expositor*, February 15, March 2, 1847); J. Palmer, *Journal of Travels over the Rocky Mountains*, 144; Ingersoll, "Overland to California," May 1, 1847; C. A. Tuttle to M. Tuttle, April 8, 1849; Delano, *California Correspondence*, April 21, 1849, 8; F. McDiarmid to C. McDiarmid, July 30, 1850; and Shepherd, "Journal of Travel," May 4, 1850, 8.

22. Mrs. Francis H. Sawyer, "Overland to California," MS C-F 103, BL; John M. McGlashan, "Overland Journal," MS Little, Box Mc, MMRL, 1; Lienhard, "California Immediately Before and After the Discovery of Gold," 3; J. Palmer, *Journal of Travels over the Rocky Mountains*, 144.

23. J. Clark, *Overland to the Gold Fields of California*, August 31, 1852, 295.

24. Unruh, *Plains Across*, 108; Merrill J. Mattes, *The Great Platte River Road: The Covered Wagon Mainline via Fort Kearny to Fort Laramie*, 39.

25. Tarbell, *Emigrant's Guide to California*, 18; Greene, *Horses at Work*, 27; A. Child, *Overland Route to California*, iv; P. L. Edwards to *St. Louis New Era*, May 25, 1843, MS 235, #007932, OHS; Marcy, *Prairie Traveler*, 28; J. Palmer, *Journal of Travels over the Rocky Mountains*, 141; Johnson and Winter, *Route Across the Rocky Mountains*, 214; Ware, *Emigrants' Guide to California*, 9n12.

26. Greene, *Horses at Work*, 27, 28; Dary, *Oregon Trail*, 79; Christian Nielsen, "Diary," May 23, 1853, MS A254, USHS; Shepherd, "Journal of Travel," 43; J. Palmer, *Journal of Travels over the Rocky Mountains*, 142; G. Belshaw, "Diary," June 17, 1853, 15.

27. S. B. Crockett to Friends, February 16, 1845, MS 1500, #007950, OHS; Ware, *Emigrants' Guide to California*, 4; Shively, *Route and Distances to Oregon and California*, 6; A. Child, *Overland Route to California*, December 10, 1850, vi; P. L. Edwards, *Sketch of the Oregon Territory; or, Emigrants' Guide* (Liberty, MO: Herald Office, 1842), PAM 297, USHS, 5; Charles Maltby to S. B. Moody, letter, "From the Plains," ca. May 24, 1849, *Illinois Daily*, ca. June 27, 1849, MS Little, MMRL; Robert L. Munkres, "Wagon Train Animals," 17.

28. Hastings, *Emigrants' Guide*, 145; Marcy, *Prairie Traveler*, 29–30; Newton Gleaves Finley, "Memories of Travel," 1852, MS C-D 5182, BL; William Henry Streeper, "Life Story of William Henry Streeper, ca. 1937," 1851, MS HM66717, HL; Kathryn Troxel, "Food of the Overland Emigrants," 21; Ware, *Emigrants' Guide to California*, 22.

29. Johnson and Winter, *Route Across the Rocky Mountains,* 212; Tarbell, *Emigrant's Guide to California,* 17; John Hauer, *The Natural Superiority of Mules,* 2; Greene, *Horses at Work,* 29; McShane and Tarr, *Horse in the City,* 13.

30. Donna Campbell Smith, *The Book of Mules: Selecting, Breeding, and Caring for Equine Hybrids,* 1–2; Homer, *The Iliad.*

31. D. Smith, *Book of Mules,* 3; Melvin Bradley, *The Missouri Mule: His Origin and Times,* 40–42; John Ashton, "History of Jack Stock and Mules in Missouri," 5.

32. Perry McCandless, *A History of Missouri,* 50; Paul C. Nagel, *Missouri: A Bicentennial History,* 60; Ashton, "History of Jack Stock and Mules in Missouri," 6, 15.

33. H. Williams, "Diary," April 19, 1859, 3; Morgan, "Diary," May 17, 1849; McGlashan, "Overland Journal," ca. April 1850; Niles Searls, *The Diary of a Pioneer, and Other Papers,* 11.

34. Tarbell, *Emigrant's Guide to California,* 17; Greene, *Horses at Work,* 5; Jack Larkin, *The Reshaping of Everyday Life, 1790–1840,* 215.

35. Hastings, *Emigrants' Guide,* 144; Shively, *Route and Distances to Oregon and California,* 3, 6; Johnson and Winter, *Route Across the Rocky Mountains,* 214; Ware, *Emigrants' Guide to California,* 9.

36. Morgan, "Diary," 1849; A. A. Enos, *Across the Plains in 1850;* McGlashan, "Overland Journal," 1850, 1.

37. Cyrus Bayles to father, April 28, 1850, MS 86/82C, BL; G. Belshaw, "Diary," April 6, 28, May 5, 7, 1853; Enos, *Across the Plains;* Batchelder, "Journal of a Tour Across the Continent," May 20, 1849, 6; F. McDiarmid to C. McDiarmid, May 12, 1850; Shepherd, "Journal of Travel," April 19, 20, 21, 1850; C. Smith, *Journal of a Trip to California,* April 25, 1850, 18; Leander V. Loomis, *A Journal of the Birmingham Emigrating Company: The Record of a Trip from Birmingham, Iowa, to Sacramento, California, in 1850,* April 28, 1850, 6.

38. F. McDiarmid to C. McDiarmid, May 22, 1850; Henry Billington Packer to Mary Judkins, April 30, 1850, MS 84/90C, BL.

39. Batchelder, "Journal of a Tour Across the Continent," May 12, 1849, 5; Morgan, "Diary," 1849.

40. Porter, "By Ox Team to California," 195; Batchelder, "Journal of a Tour Across the Continent," May 3, 16, 20, 1849, 3, 5, 6.

41. C. A. Tuttle to M. Tuttle, April 6, 1849; Batchelder, "Journal of a Tour Across the Continent," May 16, 22, 28, 1849; William Braden to Col. D. L. McFarland, April 16, 1849, MS 2006/1, BL.

42. Charles A. Brandt, "To Oregon with Ox Teams," Coll. 61, #043647, OHS.

43. August Wilhelm Wern, "Reminiscences," 1859, MS HM68166, HL; Onun Björn Dahle, "Reminiscences of Onun Björn, 1850," 11, MS Little, MMRL; Sawyer, "Overland to California," June 25, 1852; Batchelder, "Journal of a Tour Across the Continent," May 16, 1849.

44. Jean Rio Pearce, "Diary," May 22, 1851, MS 66, UU; McGlashan, "Overland Journal," 1; Batchelder, "Journal of a Tour Across the Continent," May 26, June 1, 1849; Hannah Cornaby, *Autobiography and Poems,* 32; Polly Aird, "Bound for Zion: The Ten- and Thirteen-Pound Emigrating Companies, 1853–54," 315.

45. Sandra L. Myres, *Westering Women and the Frontier Experience, 1800–1915*, xviii; Faragher, *Women and Men on the Overland Trail*, 14; Dary, *Oregon Trail*, 201, 223; Johnson and Winter, *Route Across the Rocky Mountains*, 16; Helen B. Kroll, "The Books That Enlightened the Emigrants," 106; Ray Allen Billington, "Books That Won the West," 25.

46. Hastings, *Emigrants' Guide*, 151; A. Child, *Overland Route to California*, iii; Shively, *Route and Distances to Oregon and California*, 12; Dary, *Oregon Trail*, 225; Johnson and Winter, *Route Across the Rocky Mountains*, 9.

47. Johnson and Winter, *Route Across the Rocky Mountains*, 214; Platt and Slater, *Travelers' Guide Across the Plains*, xviii.

48. Frederich Fry, *Fry's Traveler's Guide and Descriptive Journal of the Great North-Western Territories*, 9; J. Palmer, *Journal of Travels over the Rocky Mountains*, 144; Edwards to *St. Louis New Era*, May 25, 1843.

49. Johnson and Winter, *Route Across the Rocky Mountains*, 214; "From Fort Laramie," *Frontier Guardian* (Kanesville, IA), August 22, 1849; A. Child, *Overland Route to California*, vi, 56.

50. Tarbell, *Emigrant's Guide to California*, 18.

51. A. J. McCall, *The Great California Trail in 1849*, 4; Batchelder, "Journal of a Tour Across the Continent," May 3, 1849, 5; Braden to McFarland, April 16, 1849; J. Palmer, *Journal of Travels over the Rocky Mountains*, 12.

52. Hastings, *Emigrants' Guide*, 7; Marcy, *Prairie Traveler*, 67; Clark C. Spence, "Canines to Canaan," 64; Ernest S. Osgood, "Our Dog Scannon: Partner in Discovery"; Clark C. Spence, "Pioneers with Wagging Tails: Dogs on the Trail to Oregon," 27, 29; Elizabeth Wood, "Journal of a Trip to Oregon," 171–72; Solomon Tetherow, *The Organization Journal of an Emigrant Train of 1845;* Cephas Arms, "Journal Describing Overland Journey from Illinois to California," May 29, 1849, in *Knoxville (IL) Journal*, October 17, 1849, MS 67/47C, BL.

53. Sarah Watson Hamilton, *A Pioneer of Fifty-Three*, 30; Harriet Scott Palmer, "Crossing over the Great Plains by Ox-Wagons," MS 1508, #006170, OHS.

54. G. Belshaw, "Diary," May 13, 14, 1853; Frizzell, *Across the Plains to California*, May 1, 1852, 10; Eliza Ann McAuley Egbert, "Mother's Diary: The Record of a Journey Across the Plains in '52," May 10, 1852, MS 645, CHS; W. Brown, *Authentic Wagon Train Journal*, May 15, 1853, 17.

55. Dary, *Oregon Trail*, 204; W. Brown, *Authentic Wagon Train Journal*, May 19, 1853, 19; Willoughby, *Robidoux's Town*, 27.

56. H. Bradley and N. J. Bradley, "Diary," May 10, 1852; C. Smith, *Journal of a Trip to California*, April 23, 1850, 17; J. Clark, *Overland to the Gold Fields of California*, May 6, 1852, 232; Sarah Royce, *Across the Plains: Sarah Royce's Western Narrative*, 33; W. Brown, *Authentic Wagon Train Journal*, May 19, 1853, 19.

57. J. Clark, *Overland to the Gold Fields of California*, May 6, 1852, 232–33.

58. Rosemary Gudmundson Palmer, *Children's Voices from the Trail: Narratives of the Platte River Road*, 28; White, *"It's Your Misfortune and None of My Own,"* 199.

59. Frizzell, *Across the Plains to California*, May 1, 1852, 10; Sir Walter Scott, "The Lady of the Lake."

4. On the Trails

1. B. F. Nichols, "Across the Plains in 1844: Reminiscences of Oregon," in *Laidlaw (OR) Chronicle,* 1906, c2794, vol. 1, State Historical Society of Missouri, Columbia; Daniel McLaughlin, *Sketch of a Trip from Omaha to Salmon River;* Cornaby, *Autobiography and Poems,* 32; Judson, *Pioneer's Search for an Ideal Home,* 18; Batchelder, "Journal of a Tour Across the Continent," June 1, 1849, 10 (see chap. 3, n. 12).

2. Benton, "Diary," 1850 (see chap. 1, n. 3); Marcy, *Prairie Traveler,* 44–45; Mullan, *Miners and Travelers' Guide,* 9.

3. Royce, *Across the Plains,* 30, 40; Henry T. Baldy, "Journal," May 19, 20, 1852, MS OCTA, Box 2 Ba–Bl, File Baldy, MMRL; Benton, "Diary," 1850; Hastings, *Emigrants' Guide,* 148.

4. C. A. Tuttle to M. Tuttle, June 5, 1849 (see chap. 3, n. 11); Johnson and Winter, *Route Across the Rocky Mountains,* 48; Morgan, "Diary," May 10, 1849 (see chap. 3, n. 8); Charles L. Anderson to unknown [likely Mrs. Anderson], "Diary and Letters," May 20, 1862, MS P-G266, BL.

5. Barnett, *Long Trip in a Prairie Schooner,* 31.

6. Morgan, "Diary," June 5, 1849; Marcy, *Prairie Traveler,* 24; Sawyer, "Overland to California," May 24, 1852 (see chap. 3, n. 22); Hastings, *Emigrants' Guide,* 147.

7. Marcy, *Prairie Traveler,* 66–67.

8. C. Smith, *Journal of a Trip to California,* May 5, 1850, 25; Tonna, *Kindness to Animals,* 62; Royce, *Across the Plains,* 35; Judson, *Pioneer's Search for an Ideal Home,* 48, 72.

9. Hutchings, "Diaries," June 10, 17, 1849 (see chap. 3, n. 20); J. Clark, *Overland to the Gold Fields of California,* August 29, 1852, 294; C. Smith, *Journal of a Trip to California,* May 12, 1850, 28; Finley, "Memories of Travel" (see chap. 3, n. 28).

10. F. McDiarmid to C. McDiarmid, July 30, 1850 (see chap. 3, n. 2); James S. Cowden, "Diary Kept by James S. Cowden on His Trip 'Overland' from Iowa to California in 1853 with Ox Teams and Wagons," July 31, 1853, MS 465, CHS.

11. Georgia Hughes, "Recollections," 1910, MS 2999, #022628, OHS; William H. Babcock, "Journal of the Road from Missouri River," June 19, 26, 1859, MS Little, Box Ba–Be, File Babcock, MMRL; Shepherd, "Journal of Travel," June 15, 1850, 14 (see chap. 3, n. 8); Thomas Flint, *Diary of Dr. Thomas Flint, California to Maine and Return, 1851–1855,* July 23, 1853, 34.

12. Judson, *Pioneer's Search for an Ideal Home,* 36.

13. Cowden, "Diary," January 1, 1854; C. Anderson to Mrs. Charles L. Anderson, June 29, July 4, 1862; Medorem Crawford, "Journal of Medorem Crawford," June 24, 1842, 10.

14. J. Clark, *Overland to the Gold Fields of California,* May 24, 1852, 241; Royce, *Across the Plains,* 38.

15. Sawyer, "Overland to California," May 25, 1852, 6; C. Smith, *Journal of a Trip to California,* May 21, 1850, 33; Babcock, "Journal of the Road," June 2, 1859; Johnson and Winter, *Route Across the Rocky Mountains,* 52.

16. J. Clark, *Overland to the Gold Fields of California,* May 25, 26, 1852, 241–42.

17. [Mrs.] M. A. Looney, "A Trip Across the Plain in 1853," MS 281, #007867, OHS; Enos, *Across the Plains,* May 6, 1850.

18. J. Palmer, *Journal of Travels over the Rocky Mountains,* June 7, 1845, 20; Barker, "Journal of an Overland Route to California," May 18, 1850 (see chap. 1, n. 6); William A. Empey, "The Journal of William A. Empey, 1847," May 14, 1847, 124–25; Shepherd, "Journal of Travel," May 23, 1850, 7.

19. Richard N. Ellis, "Fort Kearny," 384; Platt and Slater, *Travelers' Guide Across the Plains,* 4; J. Clark, *Overland to the Gold Fields of California,* May 27, 1852, 243; McGlashan, "Overland Journal," May 3, 1850, 5 (see chap. 2, n. 22).

20. F. McDiarmid to C. McDiarmid, June 8, 1850; Sawyer, "Overland to California," May 29, 1852, 8; H. Williams, "Diary," May 11, 1859, 10 (see chap. 3, n. 3); C. Smith, *Journal of a Trip to California,* May 11, 1850, 27.

21. Judson, *Pioneer's Search for an Ideal Home,* 31; J. D. Breyfogle, *Diary of J. D. Breyfogle, Sr.: Covering His Experiences During His Overland Trip to California During the Gold Rush in 1849,* May 12, 13, 1849; J. Clark, *Overland to the Gold Fields of California,* May 27, 1852, 243; C. Smith, *Journal of a Trip to California,* May 19, 1850, 32.

22. C. Smith, *Journal of a Trip to California,* May 19, 1850, 31; J. Clark, *Overland to the Gold Fields of California,* June 2, 4, 1852, 246, 247; Barker, "Journal of an Overland Route to California," May 24, 1850; Ingersoll, "Overland to California," June 15, 1947, 21 (see chap. 3, n. 13); Richard Thomas Ackley, "A Trip Across the Plains in 1858," 12, MS A11, USHS; Judson, *Pioneer's Search for an Ideal Home,* 34.

23. Platt and Slater, *Travelers' Guide Across the Plains,* 7; Flint, "Diary," July 9, 1853, 32; C. Smith, *Journal of a Trip to California,* May 24, 1850, 36.

24. Mattes, *Great Platte River Road,* 379; Crockett to Friends, February 16, 1845 (see chap. 3, n. 27); Flint, "Diary," July 10, 1853, 32; Johnson and Winter, *Route Across the Rocky Mountains,* 56–57; Judson, *Pioneer's Search for an Ideal Home,* 36.

25. Johnson and Winter, *Route Across the Rocky Mountains,* 57–58; Dary, *Oregon Trail,* 98; A. Child, *Overland Route to California,* 57.

26. Gordon B. Dodds and Richard N. Ellis, "Fort Laramie," 384–85; Parkman, *The Oregon Trail,* 83;

27. W. Brown, *Authentic Wagon Train Journal,* June 27, 1853; Flint, "Diary," July 16, 1853, 33; F. McDiarmid to C. McDiarmid, June 20, 1850.

28. Arms, "Journal Describing Overland Journey," June 28, 1849 (see chap. 3, n. 52); McLaughlin, *Sketch of a Trip,* 4.

29. Shepherd, "Journal of Travel," June 11, 1850, 12; C. Smith, *Journal of a Trip to California,* May 29, 30, 1850, 43; Loomis, *Journal of the Birmingham Emigrating Company,* June 2, 1850, 34; Cowden, "Diary," July 2, 1853; J. R. Starr, "Diary," June 25, 1850, 20–21, MS 2473, OHS.

30. J. Clark, *Overland to the Gold Fields of California,* June 13, 1852, 252.

31. Royce, *Across the Plains,* July 1849, 43; Judson, *Pioneer's Search for an Ideal Home,* 38, 41.

32. Shepherd, "Journal of Travel," June 12, 1850, 12; Batchelder, "Journal of a Tour Across the Continent," July 21, 1849, 44.

33. Shepherd, "Journal of Travel," June 12, 1850, 13; Judson, *Pioneer's Search for an Ideal Home,* 41; Arms, "Journal Describing Overland Journey," July 13, 1849; Hutchings, "Diaries," August 4, 1849.

34. Arms, "Journal Describing Overland Journey," July 4, 1849; Royce, *Across the Plains,* July 17, 1849, 43–44.

35. Wm. H. Frush, *The Overland Diary, May 15–September 25, 1850,* June 28, 1850, 16; William Clayton, *The Latter-day Saints' Emigrants' Guide: Being a Table of Distances...from Council Bluffs to the Valley of the Great Salt Lake,* 160; Barker, "Journal of an Overland Route to California," June 10, 1850.

36. Johnson and Winter, *Route Across the Rocky Mountains,* 61; Royce, *Across the Plains,* 44; Egbert, "Mother's Diary," June 25, 1852 (see chap. 3, n. 54); M. Belshaw, *Crossing the Plains,* June 30, 1853, 27; Sawyer, "Overland to California," June 20, 1852, 13.

37. Judson, *Pioneer's Search for an Ideal Home,* 43; Royce, *Across the Plains,* 43.

38. Judson, *Pioneer's Search for an Ideal Home,* 42; John H. Benson, "From St. Joseph to Sacramento by a Forty-Niner," July 1, 1849, MS 128, UU.

39. George Washington Gill Averett, "Autobiography," MS A69, USHS; Clayton, *Latter-day Saints' Emigrants' Guide,* 160.

40. Arms, "Journal Describing Overland Journey," July 15, 1849; C. Smith, *Journal of a Trip to California,* June 12, 1850, 51–52; G. Belshaw, "Diary," July 1, 4, 1853; Babcock, "Journal of the Road," June 27, 1859; Barnett, *Long Trip in a Prairie Schooner,* 57.

41. Arms, "Journal Describing Overland Journey," July 25, 1849; Frizzell, *Across the Plains to California,* June 26, 1852, 30.

42. Judson, *Pioneer's Search for an Ideal Home,* July 12, 1853, 45–46; G. Belshaw, "Diary," July 11, 1853, 23.

43. G. Belshaw, "Diary," July 9, 1853; C. Smith, *Journal of a Trip to California,* June 16, 1850, 54; Ware, *Emigrants' Guide to California,* 25; Shepherd, "Journal of Travel," June 22, 23, 1850.

44. Judson, *Pioneer's Search for an Ideal Home,* July 12, 1853, 47.

45. Babcock, "Journal of the Road," July 10, 1859; C. Smith, *Journal of a Trip to California,* June 17, 1850, 54.

46. Judson, *Pioneer's Search for an Ideal Home,* July 12, 1853, 47; C. Smith, *Journal of a Trip to California,* June 18, 1850, 56; Sawyer, "Overland to California," June 28, 1852, 16.

47. Searls, *Diary of a Pioneer,* May 21, 1849, 18; Quincy A. Brooks to Ruter, November 7, 1851, MS 1508, OHS; F. McDiarmid to C. McDiarmid, June 28, 1850; Sawyer, "Overland to California," June 22, 1852, 14.

48. G. Belshaw, "Diary," June 14, 1853, 14; F. McDiarmid to C. McDiarmid, June 17, 20, 1850; J. Clark, *Overland to the Gold Fields of California,* 18; Brandt, "To Oregon with Ox Teams," May 20, 1851 (see chap. 3, n. 42); Nielsen, "Diary," July 13, 1853, 19 (see chap. 3, n. 26).

49. Johnson and Winter, *Route Across the Rocky Mountains,* 47; Searls, *Diary of a Pioneer,* June 8, 1849, 29; Batchelder, "Journal of a Tour Across the Continent," June 20, 1849, 20; Morgan, "Diary," June 24, 1849.

50. Ingersoll, "Overland to California," June 27, 1847; Mary E. Brush to parents, January 27, 1850, OCTA-MS, Box 3, Bo–Bu, File: Brush, MMRL; McGlashan, "Overland Journal," April 27, 1850.

51. G. Belshaw, "Diary," June 8, 1853; Royce, *Across the Plains,* 41; J. Palmer, *Journal of Travels over the Rocky Mountains,* May 12, 22, 23, 1845, 15, 18.

52. Marcy, *Prairie Traveler,* 68–69; Arms, "Journal Describing Overland Journey," May 29, June 4, 1849; Hamilton Scott, "A Trip Across the Plains in 1862," June 15, August 2, 1862, O-B-S424, Multnomah County Central Library, Portland, OR; G. Belshaw, "Diary," April 4, 1853.

53. Royce, *Across the Plains,* 31; Nielsen, "Diary," June 1, 1853; Shepherd, "Journal of Travel," May 13, 1850.

54. Benson, "From St. Joseph to Sacramento," June 25, 1849; F. McDiarmid to C. McDiarmid, August 29, 1850; Loomis, *Journal of the Birmingham Emigrating Company,* August 9, 1850, 119; Egbert, "Mother's Diary," June 4, 1852; J. Palmer, *Journal of Travels over the Rocky Mountains,* June 29, 1845, 28.

55. Nielsen, "Diary," June 3, 1853, 13; Arms, "Journal Describing Overland Journey," July 13, 19, 1849; Benson, "From St. Joseph to Sacramento," May 20, 1849.

56. J. Clark, *Overland to the Gold Fields of California,* June 9, 1852, 249; R. Ackley, "Trip Across the Plains in 1858," June 1858, 2.

57. Ossian F. Taylor Papers, "Diary," July 26, 1851, Accn. 1782, Box 1, Folder 1, UU; J. Palmer, *Journal of Travels over the Rocky Mountains,* 12.

58. Flint, "Diary," July 13, 1853, 33; Babcock, "Journal of the Road," June 21, 1859; Egbert, "Mother's Diary," July 10, 1852; Baldy, "Journal," July 19, August 27, 1852; G. Belshaw, "Diary," June 19, 21, 1853.

59. Babcock, "Journal of the Road," June 17, 1859; Frizzell, *Across the Plains to California,* June 16, 1852, 25; Loomis, *Journal of the Birmingham Emigrating Company,* June 27, 1850, 62; Ezekial W. Headley, "Book Containing Notes of Travel Across the Plains," June 14, 1849, MS 83/4C, BL; McCall, *Great California Trail,* July 25, 1849, 60; Hutchings, "Diaries," July 21, 1849.

60. Frizzell, *Across the Plains to California,* June 21, 1852, 27; Pearce, "Diary," July 26, August 16, 1851, 42, 45 (see chap. 3, n. 44).

61. Unruh, *Plains Across,* 183; McLaughlin, *Sketch of a Trip,* 4; Batchelder, "Journal of a Tour Across the Continent," May 28, 1849; Ware, *Emigrants' Guide to California,* 15, 33; A. Child, *Overland Route to California,* 46; Sawyer, "Overland to California," 20; G. Belshaw, "Diary," May 16, 1853, 8; Egbert, "Mother's Diary," May 21, 1852.

62. Looney, "Trip Across the Plain in 1853"; James O. Rayner, "Journal, 1847–1849," October 2, 1847, MS 1508, #007800, OHS; Batchelder, "Journal of a Tour Across the Continent," June 6, 1849, 13.

63. G. Belshaw, "Diary," June 19, July 21, September 1, 1853, 17, 28, 44; Barker, "Journal of an Overland Route to California," February 14, 1851; F. McDiarmid to C. McDiarmid, July 21, September 10, 1850.

64. Alexander, "Little Story," 7 (see chap. 1, n. 1).

65. Frizzell, *Across the Plains to California,* May 10, 1852, 12; Arms, "Journal Describing Overland Journey," June 15, 1849; H. Bradley and N. J. Bradley, "Diary," May 12, 1852 (see chap. 3, n. 15).

66. Batchelder, "Journal of a Tour Across the Continent," June 17, July 8, 1849, 18, 36; Thissell, *Crossing the Plains in '49,* June 20, 1849, 65.

67. Brush to parents, January 27, 1850; Batchelder, "Journal of a Tour Across the Continent," July 23, 1849, 47.

68. J. Clark, *Overland to the Gold Fields of California,* July 7, 1852, 268; Gordon B. Dodds, "Fort Bridger," 382.

69. Ware, *Emigrants' Guide to California,* 28; Batchelder, "Journal of a Tour Across the Continent," August 12, 1849; Bennett Clark, "Diary of a Journey from Missouri to California in 1849," July 4, 5, 1849, 26; Egbert, "Mother's Diary," July 13, 1852; G. Belshaw, "Diary," July 20, 1853.

70. Loomis, *Journal of the Birmingham Emigrating Company,* June 28, 1850, 63; Crawford, "Journal of Medorem Crawford," August 11, 1842, 15; Sawyer, "Overland to California," July 4, 1852, 18; A. Child, *Overland Route to California,* 36; G. Belshaw, "Diary," July 21, 1853.

71. Shepherd, "Journal of Travel," July 3, 1850, 19; Looney, "Trip Across the Plain in 1853."

72. Gordon B. Dodds, "Fort Hall," 384; J. Palmer, *Journal of Travels over the Rocky Mountains,* August 8, 1845, 42; Shively, *Route and Distances to Oregon and California,* 9; McLaughlin, *Sketch of a Trip,* 9.

73. Judson, *Pioneer's Search for an Ideal Home,* 59; Barker, "Journal of an Overland Route to California," July 2, 1850.

74. Dary, *Oregon Trail,* 107.

75. Brooks to Ruter, November 7, 1851; Royce, *Across the Plains,* June 16, 1849, 39.

76. Judson, *Pioneer's Search for an Ideal Home,* 31; Barnett, *Long Trip in a Prairie Schooner,* 60.

77. Marcy, *Prairie Traveler,* 76; Enos, *Across the Plains;* F. McDiarmid to C. McDiarmid, June 14, July 20, 1850.

78. Brandt, "To Oregon with Ox Teams," May 22, 1851; F. McDiarmid to C. McDiarmid, June 21, July 1, 1850; Loomis, *Journal of the Birmingham Emigrating Company,* July 19, 1850, 55–56.

79. William Smedley, *Across the Plains in '62,* July 8, 1862; Frizzell, *Across the Plains to California,* May 27, 1852, 20.

80. Barnett, *Long Trip in a Prairie Schooner,* 62.

81. Marcy, *Prairie Traveler,* 75; Batchelder, "Journal of a Tour Across the Continent," June 28, August 15, 1849, 26, 66.

82. Judson, *Pioneer's Search for an Ideal Home,* 41, 47; B. Clark, "Diary of a Journey from Missouri to California," June 30, 1849, 25; C. Smith, *Journal of a Trip to California,* June 6, 1850, 48.

83. See, for example, Frizzell, *Across the Plains to California,* June 18, 1852, 26; W. Brown, *Authentic Wagon Train Journal,* May 22, 1853, 21; H. Williams, "Diary," June 19, 1859, 20; Headley, "Book Containing Notes of Travel," July 3, 1849; and Shepherd, "Journal of Travel," June 24, 1850, 17.

84. Enos, *Across the Plains;* J. Clark, *Overland to the Gold Fields of California,* May 9, 1852, 235; Shepherd, "Journal of Travel," June 11, 1850, 12.

85. F. McDiarmid to C. McDiarmid, July 1, 1850; Looney, "Trip Across the Plain in 1853"; Babcock, "Journal of the Road," May 17, 1859.

86. Loomis, *Journal of the Birmingham Emigrating Company,* June 19, 1850, 55; Looney, "Trip Across the Plain in 1853"; Elizabeth Julia Goltra, *Journal Kept by Mrs. E. J. Goltra of Her Travels Across the Plains in the Year 1853,* 10; Frizzell, *Across the Plains to California,* June 18, 1852, 26.

87. Starr, "Diary," August 3, 1850; Sawyer, "Overland to California," July 3, 1852, 17; H. Williams, "Diary," June 23, 1859, 20.

88. Judson, *Pioneer's Search for an Ideal Home,* 58, 62; Starr, "Diary," August 3, 1850; Loomis, *Journal of the Birmingham Emigrating Company,* 73; Goltra, *Journal Kept by Mrs. E. J. Goltra,* August 23, 1853, 24.

89. Judson, *Pioneer's Search for an Ideal Home,* 64, 68; M. Belshaw, *Crossing the Plains,* September 3, 1853, 43; J. Palmer, *Journal of Travels over the Rocky Mountains,* September 13, 1845, 55; G. Belshaw, "Diary," September 3, 1853, 45.

90. Crockett to Friends, February 16, 1845; Stuart Richey to James Richey, November 26, 1853, Richey Family Papers, MS 1508, #007804, OHS; G. Belshaw, "Diary," September 23, 1853; Judson, *Pioneer's Search for an Ideal Home,* 71.

91. G. Belshaw, "Diary," September 15, 16, 1853; Judson, *Pioneer's Search for an Ideal Home,* 71–72.

92. Dary, *Oregon Trail,* 149; Harden, letters, September 17, 1847 (see chap. 1, n. 6); J. Palmer, *Journal of Travels over the Rocky Mountains,* 8; Looney, "Trip Across the Plain in 1853."

93. F. McDiarmid to C. McDiarmid, July 28, 29, 1850; Nielsen, "Diary," September 30, October 2, 1853.

94. A. Child, *Overland Route to California,* 59; Royce, *Across the Plains,* 47; McCall, *Great California Trail,* July 18, 25, 27, 1849, 53, 60, 61; F. McDiarmid to C. McDiarmid, August 1, 1850, 62.

95. James W. Hulse, *The Silver State: Nevada's Heritage Reinterpreted,* 45; Dary, *Oregon Trail,* 75; Horace Greeley, *An Overland Journey from New York to San Francisco in the Summer of 1859,* 229, 231; Shepherd, "Journal of Travel," July 31, 1850, 31.

96. Russell R. Elliott and William D. Rowley, *History of Nevada,* 47; Hulse, *Silver State,* 52; Shepherd, "Journal of Travel," July 16, 1850, 24.

97. B. Clark, "Diary of a Journey from Missouri to California," July 19, 1849, 30; Frink, "Adventures of a Party of Gold-Seekers," July 24, 1850, 124; Lucena Parsons, "An Overland Honeymoon," May 19, 1851, 286; Mary E. Ackley, *Crossing the Plains and Early Days in California: Memories of Girlhood Days in California's Golden Age,* 29.

98. W. Brown, *Authentic Wagon Train Journal,* August 20, September 6, 7, 8, 1853, 65, 71, 72; Sawyer, "Overland to California," July 18, 1852; William Williamson to John Williamson, November 25, 1856, T.2680, Public Record Office of Northern Ireland, Belfast; Baldy, "Journal," August 28, 29, 1852; William H. Cureton, "Trekking to California," MS C-D 5102, BL; A. Child, *Overland Route to California,* 49.

99. Marcy, *Prairie Traveler,* 52; Shepherd, "Journal of Travel," July 16, 1850, 23; Parsons, "An Overland Honeymoon," May 20, 1851, 286; Barker, "Journal of an Overland Route to California," July 19, 1850.

100. Marcy, *Prairie Traveler,* 52; Sawyer, "Overland to California," August 2, 1852, 25.

101. Breyfogle, *Diary of J. D. Breyfogle,* July 26, 29, 1849; Hutchings, "Diaries," September 27, 1849; Mary Rockwood Powers, *A Woman's Overland Journal to California,* 56, 57.

102. Morgan, "Diary," July 17, 1849; Joseph Goldsborough Bruff, "Journal, 1849–1850," September 3, 1849, MS HM 252, HL.

103. J. Clark, *Overland to the Gold Fields of California,* August 18, 1852, 287; Bruff, "Journal, 1849–1850," September 21, 1849; F. McDiarmid to C. McDiarmid, September 11, 1850; Lienhard, "California Immediately Before and After the Discovery of Gold," 146 (see chap. 3, n. 13); Tosten Kittelsen Stabaek, "An Account of a Journey to California in 1852," 117.

104. W. Williamson to J. Williamson, November 25, 1856; Frink, "Adventures of a Party of Gold-Seekers," 139; Thissell, *Crossing the Plains in '49,* 134.

105. F. McDiarmid to C. McDiarmid, September 7, 10, 1850; Bruff, "Journal, 1849–1850," September 20, 1849; J. Clark, *Overland to the Gold Fields of California,* 40.

106. Shepherd, "Journal of Travel," August 3, 1850, 33; Parsons, "An Overland Honeymoon," May 24, 1851, 288; H. Williams, "Diary," August 3, 1859, 31.

107. H. Williams, "Diary," July 29, 1859, 30; A. Child, *Overland Route to California,* 52; Powers, *Woman's Overland Journal,* 56; Baldy, "Journal," August 31, 1852.

108. Powers, *Woman's Overland Journal,* 55.

109. C. Smith, *Journal of a Trip to California,* July 18, 1850, 75, 77; Benson, "From St. Joseph to Sacramento," September 6, 1849.

110. Egbert, "Mother's Diary," September 5, 1852; Sallie Hester, "The Diary of a Pioneer Girl," September 4, 1849, 241; Joseph N. Burroughs, "Reminiscence," MS 269, CHS; Hutchings, "Diaries," October 3, 1849.

111. Barker, "Journal of an Overland Route to California," July 23, 1851; Morgan, "Diary," July 17, 1849.

112. Barker, "Journal of an Overland Route to California," July 23, 1850; Frink, "Adventures of a Party of Gold-Seekers," August 16, 1850, 139; Shepherd, "Journal of Travel," August 3, 1850, 33; Sawyer, "Overland to California," August 2, 1852, 26; John Phillip Reid, *Law for the Elephant: Property and Social Behavior on the Overland Trail,* 294–95.

113. F. McDiarmid to C. McDiarmid, September 11, 30, 1850; W. Brown, *Authentic Wagon Train Journal,* September 11, 1853, 73; Barker, "Journal of an Overland Route to California," July 19, 1850.

114. Sawyer, "Overland to California," August 3, 1852, 26; Hamden Aubrey Cagwin, "Diary," Incident 21, MS Vault 78, CHS.

115. Hugh Skinner, "Diary of 1849," August 16, 1849, 41, Box 1, Folder 2, MS 487, Fred Rosenstock Papers, UU; Breyfogle, *Diary of J. D. Breyfogle,* July 26, 1849; Enos, *Across the Plains,* 1850, letters 9 and 11.

116. Reid, *Law for the Elephant,* xi–xii; Dary, *Oregon Trail,* 210; Hester, "Diary of a Pioneer Girl," 240; Frink, "Adventures of a Party of Gold-Seekers," August 16, 1850, 140; F. McDiarmid to C. McDiarmid, September 7, 1850; Searls, *Diary of a Pioneer,* August 26, 1849, 61; B. Clark, "Diary of a Journey from Missouri to California," July 26, August 6, 1849, 34, 41.

117. Cagwin, "Diary," Incident 21; Bruff, "Journal, 1849–1850," September 20, 1849; Barker, "Journal of an Overland Route to California," July 23, 1850; Cyrus C. Loveland, *California Trail Herd: The 1850 Missouri-to-California Journal of Cyrus C. Loveland,* 110.

118. Sawyer, "Overland to California," August 7, 12, 1852.

119. M. Ackley, *Crossing the Plains,* 33; Nancy A. Hunt, "By Ox-Team to California," 323; Averett, "Autobiography"; F. McDiarmid to C. McDiarmid, September 25, 1850; W. Brown, *Authentic Wagon Train Journal,* September 24, 1853, 77; Brush to parents, January 27, 1850.

5. Caring for the Four-Footed Overlanders

1. Judson, *Pioneer's Search for an Ideal Home,* 48; McCall, *Great California Trail,* June 23, 1849, 43.

2. Loomis, *Journal of the Birmingham Emigrating Company,* July 11, 1850, 81; Shepherd, "Journal of Travel," May 23, June 10, 1850, 7, 12 (see chap. 3, n. 8).

3. Barnett, *Long Trip in a Prairie Schooner,* 109; G. Belshaw, "Diary," June 19, 1853, 17 (see chap. 3, n. 11); S. Richey to J. Richey, November 26, 1853, 4 (see chap. 4, n. 90); Bruff, "Journal, 1849–1850," September 20, 1849 (see chap. 4, n. 102).

4. Arms, "Journal Describing Overland Journey," June 25, July 4, 1849 (see chap. 3, n. 52); Batchelder, "Journal of a Tour Across the Continent," July 28, 1849, 78 (see chap. 3, n. 12); Frink, "Adventures of a Party of Gold-Seekers," 139; J. Palmer, *Journal of Travels over the Rocky Mountains,* June 15, July 17, 1845, 23, 33; Barnett, *Long Trip in a Prairie Schooner,* 111.

5. G. Belshaw, "Diary," June 8, 1853, 13; F. McDiarmid to C. McDiarmid, June 20, July 11, 1850 (see chap. 3, n. 2); Powers, *Woman's Overland Journal,* 27; H. Scott, "Trip Across the Plains," August 22, 1862.

6. Barnett, *Long Trip in a Prairie Schooner,* 59, 111; M. Ackley, *Crossing the Plains,* 33; Mary Jane Long, "Crossing the Plains in 1852," MS 1508, #002492, OHS; Powers, *Woman's Overland Journal,* 29; B. Clark, "Diary of a Journey from Missouri to California," August 6, 1849, 41.

7. J. Palmer, *Journal of Travels over the Rocky Mountains,* August 21, 1845, 46; John Spencer, "Journal," June 24, 26, 1852, MS 1508, OHS; Rachel Taylor, "Overland Trip Across the Plains," July 19, 1853, 170; F. McDiarmid to C. McDiarmid, June 26, 1850; H. Williams, "Diary," July 19, 1859, 27 (see chap. 3, n. 3); Batchelder, "Journal of a Tour Across the Continent," July 28, 1849, 78; Hastings, *Emigrants' Guide,* 145; Shively, *Route and Distances to Oregon and California,* 4; J. Clark, *Overland to the Gold Fields of California,* August 14, 1852, 283.

8. Barnett, *Long Trip in a Prairie Schooner,* 48–53; Frink, "Adventures of a Party of Gold-Seekers," 153–54; H. Palmer, "Crossing over the Great Plains by Ox-Wagons," 7–8 (see chap. 3, n. 53).

9. Margaret White Chambers, "Reminiscences," 18, MS 1508, #006154, OHS; Isaac Foster to family, July 1, 1849, in *The Foster Family: California Pioneers,* 37; Frizzell, *Across the Plains to California,* May 28, 1852, 20.

10. Searls, *Diary of a Pioneer,* June 22, 1849, 35; Shepherd, "Journal of Travel," July 28, 1850, 29; F. McDiarmid to C. McDiarmid, July 11, 1850; Amelia Knight, "Iowa to the Columbia River," August 12, 1853, 65; Hester, "Diary of a Pioneer Girl," September 4, 1849, 240.

11. Hutchings, "Diaries," October 3, 1849 (see chap. 3, n. 20); McCall, *Great California Trail,* July 3, 1849, 47; Powers, *Woman's Overland Journal,* 33; J. Clark, *Overland to the Gold Fields of California,* June 30, 1852, 265.

12. Batchelder, "Journal of a Tour Across the Continent," July 21, September 21, 1849, 44, 98; McGlashan, "Overland Journal," June 28, 1850 (see chap. 3, n. 22); Bruff, "Journal, 1849–1850," September 21, 1849; Frizzell, *Across the Plains to California,* May 14, 1852, 15.

13. John R. Shinn, "Crossing the Plains to California, 1850," July 12, 13, 1850, MS A492, USHS; Benson, "From St. Joseph to Sacramento," June–July 1849 (see chap. 4, n. 38); F. McDiarmid to C. McDiarmid, May 20, 1850; R. Ackley, "Trip Across the Plains in 1858," August 23, 1858, 24 (see chap. 4, n. 22).

14. William Pattison, "Diary," July 16, 1849, MS 1072, OHS; Batchelder, "Journal of a Tour Across the Continent," July 18, September 15, 1849; W. Williamson to J. Williamson, November 25, 1856 (see chap. 4, n. 98); Searls, *Diary of a Pioneer,* July 10, 1849, 47; Benson, "From St. Joseph to Sacramento," July 6, 1849.

15. James Hammond, "An Account of Their Journey from Ohio to California," MS 68/168C, BL; Smedley, *Across the Plains in '62,* August 4, 1862; J. Clark, *Overland to the Gold Fields of California,* August 18, 1852, 287–88; O. Taylor Papers, "Diary," May 9, 1851, 5 (see chap. 4, n. 57).

16. Nielsen, "Diary," August 4, 1853 (see chap. 3, n. 26); F. McDiarmid to C. McDiarmid, August 25, 30, November 10, 1850; Frizzell, *Across the Plains to California,* June 23, 1852, 28.

17. Marcy, *Prairie Traveler,* 45; Mullan, *Miners and Travelers' Guide,* 9; Pearce, "Diary," May 22, 1851 (see chap. 3, n. 44); Powers, *Woman's Overland Journal,* 32; F. McDiarmid to C. McDiarmid, June 20, 1850.

18. Virginia Reed Murphy, *Across the Plains in the Donner Party: A Personal Narrative of the Overland Trip to California,* October 5, 1846, 27; Benton, "Diary," 1850 (see chap. 1, n. 3); F. McDiarmid to C. McDiarmid, July 3, 1850; Henry A. Allyn, "Journal," June 30, 1853, MS 1508, OHS; McCall, *Great California Trail;* Morgan, "Diary," ca. May 9, 1849.

19. Judson, *Pioneer's Search for an Ideal Home,* 48; Lienhard, "California Immediately Before and After the Discovery of Gold," 67 (see chap. 3, n. 13); McLaughlin, *Sketch of a Trip,* January 18, 1863, 10; Ingersoll, "Overland to California," June 29, 1847 (see chap. 3, n. 13); McCall, *Great California Trail,* June 28, 1849, 45.

20. C. Anderson to "My Own Dear Wife," May 18, June 15, 17, 1862 (see chap. 4, n. 4); O. Taylor Papers, "Diary," May 17, 1851; McCall, *Great California Trail,* June 25, 1849, 43.

21. Egbert, "Mother's Diary," April 20, June 6, 10, July 4, 21, 1852 (see chap. 3, n. 54); Thissell, *Crossing the Plains in '49,* 33, 35.

22. Rogers, *First Friend,* 108; Spence, "Canines to Canaan," 60.

23. Barnett, *Long Trip in a Prairie Schooner,* March 31, May 9, 1859, 12–13, 28, 31; Rayner, "Journal, 1847–1849," April 10, 1847 (see chap. 4, n. 62); Lienhard, "California Immediately Before and After the Discovery of Gold," 12; Thomas Jefferson Kirkpatrick, *The Kirkpatrick Story: The Day by Day Report of the Trek Across the Plains by the Kirkpatrick Party in 1854,* 3.

24. Spence, "Canines to Canaan," 62; Spence, "Pioneers with Wagging Tails," 29; Eugenia Zieber, "Journal of Our Journey to Oregon," July 4, 1851, 195; Thissell, *Crossing the Plains in '49,* May 6, 1849, 47.

25. Spence, "Canines to Canaan," 62; Barnett, *Long Trip in a Prairie Schooner,* 61, 119.

26. Spencer, "Journal," May 10, 11, 1852; Sawyer, "Overland to California," August 15, 1852, 30 (see chap. 3, n. 22); Chambers, "Reminiscences," 11.

27. Spencer, "Journal," October 24, 1852; R. Ackley, "Trip Across the Plains in 1858," June 23, 1858; Harriet A. Loughary, "Travels and Incidents, 1864," June 11, 1864, 136; H. Scott, "Trip Across the Plains," July 22, 1862; Thissell, *Crossing the Plains in '49,* 35; Spence, "Canines to Canaan," 62; Spence, "Pioneers with Wagging Tails," 30; Charles Hentz to Mother, "Letters to His Mother," May 21, 1849, MS C-B 704, BL.

28. C. Smith, *Journal of a Trip to California,* May 24, 1850, 36; Batchelder, "Journal of a Tour Across the Continent," August 17, 1849; Thissell, *Crossing the Plains in '49,* June 27, 1849, 70.

29. Bert W. Bierer, *A Short History of Veterinary Medicine in America,* 16, 90; Philip M. Teigen, "Nineteenth-Century Veterinary Medicine as an Urban Profession," n.p.; McShane and Tarr, *Horse in the City,* 157; Noah Webster, *An American Dictionary of the English Language,* Compact Edition (reprint, West Valley City, UT: Waking Lion Press, 2010), 284.

30. William Youatt, *The Horse,* 398–417; McShane and Tarr, *Horse in the City,* 158; J. F. Smithcors, *The Veterinarian in America, 1625–1975,* 72.

31. Smithcors, *Veterinarian in America,* 60–62; Fred J. Born, *The History of the Horse Doctor (*Equarius medicus*): A Brief Summary of 4000 Years of Veterinary History;* George H. Dadd, "Our Platform."

32. G. H. Dadd, *The American Cattle Doctor,* 10.

33. Hastings, *Emigrants' Guide,* 9, 146; Shively, *Route and Distances to Oregon and California,* 4; Ware, *Emigrants' Guide to California,* 9; Mullan, *Miners and Travelers' Guide,* 8.

34. Barnett, *Long Trip in a Prairie Schooner,* 42; Babcock, "Journal of the Road," June 10, 28, 1859 (see chap. 4, n. 11).

35. McShane and Tarr, *Horse in the City,* 157; F. McDiarmid to C. McDiarmid, July 3, 1850; H. Bradley and N. J. Bradley, "Diary," June 25, 1852 (see chap. 3, n. 15); G. Belshaw, "Diary," July 7, 1853; Sawyer, "Overland to California," June 13, 1852, 11; Flint, "Diary," July 11, 1853, 32 (see chap. 4, n. 11); Stabaek, "Account of a Journey," 110, 111; Loomis, *Journal of the Birmingham Emigrating Company,* June 27, 1850, 62.

36. F. McDiarmid to C. McDiarmid, May 22, 1850; J. Palmer, *Journal of Travels over the Rocky Mountains,* June 6, 1845, 20; Johnson and Winter, *Route Across the Rocky Mountains,* 52; Crockett to Friends, February 16, 1845 (see chap. 3, n. 27); Lucy Rutledge

Cooke, "Letters on the Way to California," June 10, 1852, 252; Babcock, "Journal of the Road," July 15, 1859; G. Belshaw, "Diary," June 14, 1853.

37. Brian Altonen, "Alkali Intoxication of Horses" (blog), http://brianaltonenmph.com/6-history-of-medicine-and-pharmacy; Johnson and Winter, *Route Across the Rocky Mountains*, 52n9.

38. Tarbell, *Emigrant's Guide to California*, 14; Barker, "Journal of an Overland Route to California," June 8, 1850 (see chap. 1, n. 6); G. Belshaw, "Diary," July 2, 1853, 20; Marcy, *Prairie Traveler*, 22.

39. Tarbell, *Emigrant's Guide to California*, 6; Horn, *Horn's Overland Guide to California*, 49; A. Child, *Overland Route to California*, 25; Clayton, *Latter-day Saints' Emigrants' Guide*, preface.

40. F. McDiarmid to C. McDiarmid, July 1, 2, 1850; Ingersoll, "Overland to California," July 9, 1847; G. Belshaw, "Diary," June 13, 19, 1853, 14.

41. Marcy, *Prairie Traveler*, 125; Altonen, "Alkali Intoxication of Horses"; H. Williams, "Diary," July 30, 1859, 30; Babcock, "Journal of the Road," June 4, 1859.

42. Platt and Slater, *Travelers' Guide Across the Plains*, 12, 23; Baldy, "Journal," June 30, 1852 (see chap. 4, n. 3); F. McDiarmid to C. McDiarmid, July 4, 1850; Arms, "Journal Describing Overland Journey," July 16, 1849.

43. John Ismay, "Selenium Poisoning in Horses (Alkali Disease)"; Powers, *Woman's Overland Journal*, 49; G. Belshaw, "Diary," August 25, 1853, 42; Smithcors, *Veterinarian in America*, 70.

44. Sawyer, "Overland to California," June 11, 1852, 11; G. Belshaw, "Diary," June 27, July 2, August 14, 1853, 19–20, 36; Babcock, "Journal of the Road," August 8, 10, 1859.

45. Babcock, "Journal of the Road," August 6, 1859; R. Ackley, "Trip Across the Plains in 1858," August 13, 1858, 19; Frizzell, *Across the Plains to California*, June 26, 1852, 30; Mary Ringo, "The 1864 Journal," July 25, 1864, 216; Dadd, *The American Cattle Doctor*, 189–90.

46. G. Belshaw, "Diary," August 13, 1853, 35; R. Ackley, "Trip Across the Plains in 1858," August 13, 1858, 19; Babcock, "Journal of the Road," August 6, 1859.

47. Youatt, *The Horse*, 403, 413; "Effects of Opium," 432.

48. Youatt, *The Horse*, 417; Mary Burrell, "Council Bluffs to California, 1854," August 1, 1854, 250–51.

49. Egbert, "Mother's Diary," April 30, 1852; Rayner, "Journal, 1847–1849," April 23, 1847; Marcy, *Prairie Traveler*, 124, 126.

50. G. Belshaw, "Diary," June 21, July 5, 1853, 17, 21; Hughes, "Recollections" (see chap. 4, n. 11); F. McDiarmid to C. McDiarmid, July 11, 1850; Marcy, *Prairie Traveler*, 54; J. Palmer, *Journal of Travels over the Rocky Mountains*, June 7, 1845, 20; Long, "Crossing the Plains in 1852," 9.

51. Jacqueline Williams, *Wagon Wheel Kitchens: Food on the Oregon Trail*, 70; Baldy, "Journal," June 10, 1852; Hughes, "Recollections"; Chambers, "Reminiscences," 22; Frizzell, *Across the Plains to California*, June 22, 1852, 28; Brandt, "To Oregon with Ox Teams," July 1, 1851 (see chap. 3, n. 42).

52. Egbert, "Mother's Diary," June 10, 27, 1852; Goltra, *Journal*, July 24, 25, 1853, 17; Benton, "Diary," 1850; H. Bradley and N. J. Bradley, "Diary," June 12, 1852; W. Brown, *Authentic Wagon Train Journal*, July 19, 1853, 52; Barker, "Journal of an Overland Route to California," June 13, 22, 1850; Altonen, "Alkali Intoxication of Horses"; Marcy, *Prairie Traveler*, 125.

53. H. Williams, "Diary," July 5, August 1, 1859, 24, 31; Barnett, *Long Trip in a Prairie Schooner*, 31; Loomis, *Journal of the Birmingham Emigrating Company*, 14, 43; Batchelder, "Journal of a Tour Across the Continent," July 19, 1849, 42; G. Belshaw, "Diary," August 13, 1853; Knight, "Iowa to the Columbia River," August 1, 1853, 62; Pearce, "Diary," July 24, 1851; M. Ackley, *Crossing the Plains*, 31; Hamilton, *Pioneer of Fifty-Three*, 36.

54. Shinn, "Crossing the Plains to California," October 1, 1850; Loomis, *Journal of the Birmingham Emigrating Company*, August 15, 1850, 127; J. Palmer, *Journal of Travels over the Rocky Mountains*, September 21, 1845; W. Williamson to J. Williamson, November 25, 1856.

55. Johnson and Winter, *Route Across the Rocky Mountains*, 81; Crockett to Friends, February 16, 1845; Barker, "Journal of an Overland Route to California," July 19, 1850; Loomis, *Journal of the Birmingham Emigrating Company*, July 30, 31, 1850, 110–11.

56. H. Williams, "Diary," July 6, 1859, 24; Benton, "Diary," 1850.

57. W. Williamson to J. Williamson, November 25, 1856; Thissell, *Crossing the Plains in '49*, September 16, 1849, 173; Peter Frederick Hummel to "Dear Wife and Children," April 27, 1850, MS HM 70759, HL; Shepherd, "Journal of Travel," August 24, 25, 1850, 42.

58. Cagwin, "Diary," October 20, 1850 (see chap. 4, n. 114); Shepherd, "Journal of Travel," August 15, 1850, 39; John T. Gridley to Elisha Gridley, August 4, 1850, George Gridley Papers, MS 82/62C, BL; Sawyer, "Overland to California," August 15, 1852, 30; Enos, *Across the Plains*, July 1850; Barker, "Journal of an Overland Route to California," August 3, 1850.

59. Loomis, *Journal of the Birmingham Emigrating Company*, August 13, 15, 1850, 124, 125, 129, 130.

60. Hamilton, *Pioneer of Fifty-Three*, 28, 72; Buchanan, "Letter and Circular of Information," 310 (February 15, 1847); S. Richey to John Cunningham, January 8, 1854; Isaac Julian Harvey, "To California in 1850," MS C-D 5091, BL, University of California–Berkeley.

6. Encounters with Wild Animals

1. William Cronon, *Changes in the Land: Indians, Colonists, and the Ecology of New England*, 108.

2. J. Palmer, *Journal of Travels over the Rocky Mountains*, July 29, September 11, 1845, 36, 53; Platt and Slater, *Travelers' Guide Across the Plains*, 3, 5, 9; A. Child, *Overland Route to California*, 10, 15.

3. Hentz to Mother, "Letters to His Mother," April 21, 1849 (see chap. 5, n. 27); Barker, "Journal of an Overland Route to California," May 16, 1850 (see chap. 1, n. 6);

Packer to Mary Judkins, May 28, 1850 (see chap. 3, n. 38); Elias Daggy, "Journal," June 6, 1850, MS 4079, CHS.

4. Lewis Stout, "Diary," June 9, 1852, MS 1059, OHS; Averett, "Autobiography" (see chap. 4, n. 39); Bruff, "Journal, 1849–1850," September 10, 1849 (see chap. 4, n. 102); Nielsen, "Diary," August 2, 1853, 23 (see chap. 3, n. 26).

5. William L. Cole, *California: Its Scenery, Climate, Productions and Inhabitants,* 13; Breyfogle, *Diary of J. D. Breyfogle,* May 9.

6. Jacob Hammer, *This Emigrating Company: The 1844 Oregon Trail Journal of Jacob Hammer,* June 14, 18, 29, 83, 84, 88; Cagwin, "Diary," May 16, 21, 1850 (see chap. 4, n. 114); Baldy, "Journal," May 29, June 4, 5, 11, 24, July 4, 12, 19, 30, August 7 (see chap. 4, n. 3); W. Brown, *Authentic Wagon Train Journal,* May 30, 31, 1853; Barnett, *Long Trip in a Prairie Schooner,* 20, 21, 37, 43, 44, 49, 64; Frizzell, *Across the Plains to California,* May 29, 1852, 21.

7. Hastings, *Emigrants' Guide,* 8, 135; Sophia L. Goodridge Hardy, "Journal," July 18, 1850, MS A583, USHS; Breyfogle, *Diary of J. D. Breyfogle,* May 10, 1849.

8. David A. Dary, *The Buffalo Book: The Full Saga of the American Animal,* 7–13, 18–19, 21; David A. Dary, *Buffalo,* 139–40; Dan Flores, "Bison Ecology and Bison Diplomacy: The Southern Plains from 1800 to 1850," 481.

9. William Wilson Cowan to Phebe Ann and Mary Jane Cowan, 1849, Item 8: HM68068, HL; Hardy, "Journal," July 30, 1850; Murphy, *Across the Plains in the Donner Party,* 17.

10. Brandt, "To Oregon with Ox Teams," June 8, 1851 (see chap. 3, n. 42); Sarah J. Cummins, "Autobiography and Reminiscences," PAM 11163, 1845, 24, USHS; John B. Wyeth, *Oregon,* 30; Levi Jackman, "Journal," May 6, 1847, MS A1–2, USHS; Alexander, "Little Story," May 19, 1859, 31 (see chap. 1, n. 1).

11. J. Palmer, *Journal of Travels over the Rocky Mountains,* July 9, 1845, 31; Loomis, *Journal of the Birmingham Emigrating Company,* May 22, 1850, 19; Empey, "Journal of William A. Empey," May 14, 1847; Norton Jacob, "Daily Journal," May 8, 1847, MS A29, USHS.

12. C. Smith, *Journal of a Trip to California,* May 19, 1850, 31; Cummins, "Autobiography and Reminiscences," 24; Averett, "Autobiography"; "A Ride on a Buffalo," *Nautical Standard and Steam Navigation Gazette* (London), August 9, 1851.

13. Pearce, "Diary," July 30, 1851, 43 (see chap. 3, n. 44); Jackman, "Journal," May 7, 1847; Jacob, "Daily Journal," May 5, 1847, 65; Lienhard, "California Immediately Before and After the Discovery of Gold," 29 (see chap. 3, n. 13).

14. Wyeth, *Oregon,* 29; Egbert, "Mother's Diary," June 1, 1852 (see chap. 3, n. 54); Cagwin, "Diary," May 29, 1850; J. Clark, *Overland to the Gold Fields of California,* 245; M. Ackley, *Crossing the Plains,* July 1858, 10; W. W. Cowan to P. A. and M. J. Cowan, July 30, 1849; Ruth Barnes Moynihan, "Children and Young People on the Overland Trail," 284.

15. Judson, *Pioneer's Search for an Ideal Home,* 37–38; Hutchings, "Diaries," August 22, 1848 (see chap. 3, n. 20); Crawford, "Journal of Medorem Crawford," July 14, 26, 1842, 12–13; J. Clark, *Overland to the Gold Fields of California,* 245; Frizzell, *Across the Plains to California,* June 23, 1852, 28.

16. Lienhard, "California Immediately Before and After the Discovery of Gold," 49; "Prairie Dogs and Their Cities," *Watertown (WI) Chronicle,* May 16, 1849; "Life in Nebraska," *Sauk County (WI) Standard,* May 10, 1854; Flint, "Diary," June 23, 1853 (see chap. 4, n. 11); Foster to family, May 31, 1849, in *Foster Family,* 27.

17. Russell A. Graves, *The Prairie Dog: Sentinel of the Plains,* 6–7, 10–11; Alexander R. L. Cayton, "Lewis and Clark Expedition," 397.

18. Batchelder, "Journal of a Tour Across the Continent," June 29, 1849 (see chap. 3, n. 12); Graves, *Prairie Dog,* 12; Starr, "Diary," May 31, 1850, 10 (see chap. 4, n. 29); W. Brown, *Authentic Wagon Train Journal,* June 5, 1853, 27; Cureton, "Trekking to California," 50 (see chap. 4, n. 98); J. Palmer, *Journal of Travels over the Rocky Mountains,* June 10, 1845, 21; "Prairie Dogs and Their Cities."

19. Frizzell, *Across the Plains to California,* May 29, 1852, 21; Lienhard, "California Immediately Before and After the Discovery of Gold," 49; Arms, "Journal Describing Overland Journey," June 7, 1849 (see chap. 3, n. 52); Washington Irving, *A Tour of the Prairies,* 195; Baldy, "Journal," June 4, 1852; Johnson and Winter, *Route Across the Rocky Mountains,* 52; Hammer, *This Emigrating Company,* June 14, 1844, 83; Porter, "By Ox Team to California," 234.

20. Nichols, "Across the Plains in 1844" (see chap. 4, n. 1); Irving, *Tour of the Prairies,* 194; Emeline J. Fuller, "Left by the Indians," 10; Frizzell, *Across the Plains to California,* June 8, 1852.

21. Egbert, "Mother's Diary," May 27, 1852; Searls, *Diary of a Pioneer,* June 10, 1849, 30; Thissell, *Crossing the Plains in '49,* May 16, 1849, 53.

22. Carl B. Koford, "Prairie Dogs," 908; Graves, *Prairie Dog,* 109; J. Palmer, *Journal of Travels over the Rocky Mountains,* June 10, 1845, 21; Foster to family, May 31, 1849, in *Foster Family,* 27; William Dresser to Sarah Dresser, May 29, 1850, Dresser Family Papers, MS 69/115C, BL; M. Ackley, *Crossing the Plains,* 24.

23. C. Smith, *Journal of a Trip to California,* April 40, 1850, 21; Henrietta Catherine Ellis and James Addison Ellis, *Across the West by Covered Wagon, 1865,* June 2, 1865, 17; Finley, "Memories of Travel" (see chap. 3, n. 28).

24. Cagwin, "Diary," June 6, 1850; Cureton, "Trekking to California," 50; Egbert, "Mother's Diary," June 9, 1852.

25. Starr, "Diary," May 31, June 1, 1850, 10–11; Frizzell, *Across the Plains to California,* May 29, 1852, 21; Foster to family, May 30, 1849, in *Foster Family,* 27; Fuller, "Left by the Indians," 10.

26. Scott E. Hygnstrom, "Black-Tailed Prairie Dog," 623; Irving, *Tour of the Prairies,* 195; Hamilton, *Pioneer of Fifty-Three,* 60; Batchelder, "Journal of a Tour Across the Continent," June 29, 1849; Parkman, *The Oregon Trail,* 67; Cureton, "Trekking to California," 50.

27. McCall, *Great California Trail,* 28; "The Prairie Dog," *Southport Telegraph* (Kenosha, WI), August 31, 1842; Irving, *Tour of the Prairies,* 194; J. Palmer, *Journal of Travels over the Rocky Mountains,* June 10, 1845, 21; Smedley, *Across the Plains in '62,* June 12, 1862; Brandt, "To Oregon with Ox Teams," June 7, 1851.

28. Irving, *Tour of the Prairies,* 198; "Prairie Dogs and Their Cities"; McCall, *Great California Trail,* 28; "The Prairie Dog."

29. John L. O'Sullivan, "Annexation," 5.

30. Irving, *Tour of the Prairies,* 198.

31. Smedley, *Across the Plains in '62,* August 4, 1862; Flint, "Diary," July 2, 3, 4, 5, 1853; California Wolf Center, "Comparing Wolves and Coyotes"; Emily Smith, "Wolves in the Wild," 31; Loomis, *Journal of the Birmingham Emigrating Company,* May 9, 1850, 10; Brandt, "To Oregon with Ox Teams," June 4, 1851; W. Brown, *Authentic Wagon Train Journal,* June 14, 1853, 31; Lucia Darling, "Crossing the Prairies in a Covered Wagon," July 25, 1863, 17, MS Little, Box D, MMRL; Barnett, *Long Trip in a Prairie Schooner,* 1859, 31, 53.

32. Morgan, "Diary," 1849 (see chap. 3, n. 8); Lienhard, "California Immediately Before and After the Discovery of Gold," 32; Hutchings, "Diaries," June 18, 1848.

33. Empey, "Journal of William A. Empey," May 14, 1847; Lienhard, "California Immediately Before and After the Discovery of Gold," 146; Nielsen, "Diary," August 21, September 4, 13, 14, 1853, 28, 30, 31; W. Brown, *Authentic Wagon Train Journal,* June 16, 1853, 32; Pearce, "Diary," June 21, 1851, 31; H. Bradley and N. J. Bradley, "Diary," June 1, 1852 (see chap. 3, n. 15).

34. Frizzell, *Across the Plains to California,* May 19, 1852, 16; Nichols, "Across the Plains in 1844"; Louisa M. Sweetland, "Remembrances," 1863, MS 376, CHS; H. Palmer, "Crossing over the Great Plains by Ox-Wagons," 3, 5 (see chap. 3, n. 53).

35. J. Clark, *Overland to the Gold Fields of California,* May 10, 1852, 235; Hutchings, "Diaries," June 17, July 6, 30, August 1, 1848; Flint, "Diary," June 19, 1853, 27.

36. Nichols, "Across the Plains in 1844"; Batchelder, "Journal of a Tour Across the Continent," June 22, 1849; Lienhard, "California Immediately Before and After the Discovery of Gold," 146; Capt. John I. Lewis, *My Garden of Roses; or, The Footnote of Life,* 1853, 18; Baldy, "Journal," August 23, 1852; Nielsen, "Diary," August 30, September 1, 12, 1853; W. Brown, *Authentic Wagon Train Journal,* July 3, 1853, 42; Irving, *Tour of the Prairies,* 186.

37. Marion Schwartz, *A History of Dogs in the Early Americas,* 6; Spence, "Canines to Canaan," 62; Frizzell, *Across the Plains to California,* June 17, 1852, 25; Judson, *Pioneer's Search for an Ideal Home,* 33; Brandt, "To Oregon with Ox Teams," June 4, 5, 1851; Empey, "Journal of William A. Empey," May 9, 1847; Pearce, "Diary," September 11, 1851, 49.

38. Johnson and Winter, *Route Across the Rocky Mountains,* 52; G. Belshaw, "Diary," June 24, 1853 (see chap. 3, n. 11); Foster to family, June 6, 1849, in *Foster Family,* 28; Cummins, "Autobiography and Reminiscences," 25; Smedley, *Across the Plains in '62,* June 3, 1862; Hentz to Mother, "Letters to His Mother," June 1, 1849; Murphy, *Across the Plains in the Donner Party,* 17.

39. National Geographic, "Grizzly Bear"; Hutchings, "Diaries," August 7, 8, 1848; Johnson and Winter, *Route Across the Rocky Mountains,* 63; Babcock, "Journal of the Road," June 29, 1859 (see chap. 4, n. 11); Flint, "Diary," September 4, 1853, 47; John Rueger, "The Hardships Experienced by John Rueger in Crossing the Plains During the Gold Rush," MS C-D 5059, BL.

40. Wyeth, *Oregon,* 35; Darling, "Crossing the Prairies," July 7, 1863.

41. Nielsen, "Diary," August 19, 1853, 27; Kirkpatrick, *Kirkpatrick Story,* 8; Egbert, "Mother's Diary," May 25, July 22, 1852; Smedley, *Across the Plains in '62,* July 20, 1862; Baldy, "Journal," May 26, 1852.

42. Hutchings, "Diaries," July 30, 1848; H. Ellis and J. Ellis, *Across the West by Covered Wagon,* 25; C. Smith, *Journal of a Trip to California,* June 21, 1850, 58; Thissell, *Crossing the Plains in '49,* April 6, 1849, 25; Loomis, *Journal of the Birmingham Emigrating Company,* July 1, 1850, 68; Searls, *Diary of a Pioneer,* July 22, 1849, 35; Batchelder, "Journal of a Tour Across the Continent," July 7, 1849.

43. Cowden, "Diary," July 9, 1853 (see chap. 4, n. 10); W. Brown, *Authentic Wagon Train Journal,* June 21, 1853, 35; Smedley, *Across the Plains in '62,* June 23, 1862.

44. G. Belshaw, "Diary," July 17, 1853, 26; Sawyer, "Overland to California," May 26, July 22, 23, 1852, 7 (see chap. 3, n. 22); Baldy, "Journal," July 19, 1852; H. Ellis and J. Ellis, *Across the West by Covered Wagon,* July 31, 1865, 68; H. Williams, "Diary," July 25, 1859; Unruh, *Plains Across,* 408–9.

45. Smedley, *Across the Plains in '62,* July 18, 21, 1962; Sawyer, "Overland to California," July 2, 1852, 17; H. A. Stine to E. Stine, July 5, 1850 (see chap. 3, n. 19); H. Ellis and J. Ellis, *Across the West by Covered Wagon,* August 1, 1865, 41; McGlashan, "Overland Journal," June 23, 1850 (see chap. 3, n. 22).

46. F. McDiarmid to C. McDiarmid, June 15, July 19, 1850 (see chap. 3, n. 2); Barker, "Journal of an Overland Route to California," May 30, 1850; Barnett, *Long Trip in a Prairie Schooner,* 54; Babcock, "Journal of the Road," July 25, 1859; Smedley, *Across the Plains in '62,* July 12, 1862; Flint, "Diary," June 19, 1853.

47. J. Palmer, *Journal of Travels over the Rocky Mountains,* April 28, August 3, 1845, 11, 37–38; Batchelder, "Journal of a Tour Across the Continent," June 11, July 3, 1849, 15, 31; Searls, *Diary of a Pioneer,* June 4, 1849, 26; Michael W. Klaus, "Mormon Cricket"; F. McDiarmid to C. McDiarmid, June 30, 1850; Brandt, "To Oregon with Ox Teams," July 27, 1851; Hutchings, "Diaries," July 1, 1849; Nielsen, "Diary," June 11, July 7, 1853.

48. M. Ackley, *Crossing the Plains,* July, August 6, 1858, 14, 18; Hamilton, *Pioneer of Fifty-Three,* 62; Horn, *Horn's Overland Guide to California,* 16; Clayton, *Latter-day Saints' Emigrants' Guide,* 153; Batchelder, "Journal of a Tour Across the Continent," June 3, 1849; Hutchings, "Diaries," July 11, 1849; Porter, "By Ox Team to California," 225; Flint, "Diary," July 7, 1853, 31; Bruff, "Journal, 1849–1850," September 7, 1849.

49. Egbert, "Mother's Diary," June 6, 1852; Clayton, *Latter-day Saints' Emigrants' Guide,* 149; Cagwin, "Diary," June 8, 1850; Hutchings, "Diaries," July 29, 1848; Frizzell, *Across the Plains to California,* June 17, 1852, 25; Benson, "From St. Joseph to Sacramento," June 22, 1849 (see chap. 4, n. 38); Darling, "Crossing the Prairies," July 5, 1863, 8.

50. Hastings, *Emigrants' Guide,* 99; Batchelder, "Journal of a Tour Across the Continent," June 12, 23, 1849, 15, 22; Barnett, *Long Trip in a Prairie Schooner,* 37; Breyfogle, *Diary of J. D. Breyfogle,* May 13, 1849; "Sketches of the Great West," *Huron Reflector* (Norwalk, OH), October 28, 1845.

51. Starr, "Diary," May 31, 1850, 10; H. Ellis and J. Ellis, *Across the West by Covered Wagon,* June 2, 1865, 17; R. Ackley, "Trip Across the Plains in 1858," August 13, 1858, 19 (see chap. 4, n. 22); G. Belshaw, "Diary," June 2, 1853.

52. Jacob, "Daily Journal," May 5, 1847, 64–65; Baldy, "Journal," June 5, 1852.

53. "Sketches of the Great West"; Darling, "Crossing the Prairies," July 16, 1863, 12; W. Dresser to S. Dresser, May 29, 1850; Skinner, "Diary," June 25, 1849, 21 (see chap. 4, n. 115); G. Belshaw, "Diary," June 24, August 13, 1853.

54. J. Palmer, *Journal of Travels over the Rocky Mountains,* August 12, 1845, 44; Flint, "Diary," August 13, 1853, 42; Smedley, *Across the Plains in '62,* June 8, 1862.

55. Crockett to Friends, February 16, 1845 (see chap. 3, n. 27); G. Belshaw, "Diary," May 31, 1853.

56. Allen, *Allen's Guide Book,* 5; Fry, *Fry's Traveler's Guide,* 29; "Life in Nebraska."

57. Irving, *Tour of the Prairies,* 175; W. Dresser to S. Dresser, May 29, 1850; J. Palmer, *Journal of Travels over the Rocky Mountains,* July 15, 1845, 32; C. Smith, *Journal of a Trip to California,* May 19, 1850, 32; Thissell, *Crossing the Plains in '49,* May 21, 1849, 55; Cummins, "Autobiography and Reminiscences," 24.

58. W. Brown, *Authentic Wagon Train Journal,* July 1, 1853, 41; Enos, *Across the Plains,* 1850; Batchelder, "Journal of a Tour Across the Continent," July 22, 1849.

59. Averett, "Autobiography," 1859; J. Clark, *Overland to the Gold Fields of California,* 238–39.

60. Smedley, *Across the Plains in '62,* July 11, 1862; Shively, *Route and Distances to Oregon and California,* 4; McGlashan, "Overland Journal," May 9, 1850; F. McDiarmid to C. McDiarmid, June 12, 1850; Cummins, "Autobiography and Reminiscences," 26; Egbert, "Mother's Diary," June 5, 1852.

61. C. Smith, *Journal of a Trip to California,* May 21, 1850, 34; Loomis, *Journal of the Birmingham Emigrating Company,* May 15, 1850, 15; Shepherd, "Journal of Travel," May 24, 1850, 7 (see chap. 3, n. 8); Hutchings, "Diaries," July 4, 1848.

62. Babcock, "Journal of the Road," May 26, 1859; Jacob, "Daily Journal," May 5, 1847, 64–65; Jackman, "Journal," May 7, 1847; Hardy, "Journal," July 25, 1850.

63. Batchelder, "Journal of a Tour Across the Continent," July 29, 1849; Hardy, "Journal," July 20, 22, 23, August 6, 1850; Loomis, *Journal of the Birmingham Emigrating Company,* May 17, 19, 1850, 17–18; Sawyer, "Overland to California," May 17, 1852, 4; Morgan, "Diary," May 10, 1849; Shepherd, "Journal of Travel," May 25, 1850, 7; Cowden, "Diary," June 12–15, 1853; C. Smith, *Journal of a Trip to California,* May 15, 1850, 29.

64. Hamilton, *Pioneer of Fifty-Three,* 40; Crockett to Friends, February 16, 1845; C. A. Tuttle to M. Tuttle, May 16, 1849 (see chap. 3, n. 11); Breyfogle, *Diary of J. D. Breyfogle,* May 12, 1849; Batchelder, "Journal of a Tour Across the Continent," June 26, 1849, 25; J. Palmer, *Journal of Travels over the Rocky Mountains,* 143; Avery, *Land of Ophir,* 4.

65. J. Palmer, *Journal of Travels over the Rocky Mountains,* June 15, 1845, 23; Cummins, "Autobiography and Reminiscences," 25; Streeper, "Life Story" (see chap. 3, n. 28); Troxel, "Food of the Overland Emigrants," 20.

66. Batchelder, "Journal of a Tour Across the Continent," August 24, 1849; McCall, *Great California Trail,* June 17, 1849, 38; Cowden, "Diary," June 21, 1853; Jacob, "Daily Journal," April 1847, 62; Hutchings, "Diaries," August 5, 1848; Thissell, *Crossing the Plains in '49,* May 16, 1849, 53.

67. H. Williams, "Diary," June 13, 1859, 18; Barker, "Journal of an Overland Route to California," June 3, 1850; Nichols, "Across the Plains in 1844"; Cowden, "Diary," July 14, 1853; H. Scott, "Trip Across the Plains," July 15, 1862; McCall, *Great California Trail*, June 1, 1849, 28; Sawyer, "Overland to California," June 17, 1852, 12.

68. G. Belshaw, "Diary," June 5, 1853, 12; Loomis, *Journal of the Birmingham Emigrating Company*, June 24, 1850, 59; Enos, *Across the Plains*, 1850.

69. Batchelder, "Journal of a Tour Across the Continent," July 28, 1849, 50; Babcock, "Journal of the Road," June 27, 1859.

70. Morgan, "Diary," June 6, 1849; Breyfogle, *Diary of J. D. Breyfogle*, June 5, 1849; "Sketches of the Great West"; Baldy, "Journal," September 7, 1852; Parsons, "An Overland Honeymoon," September 2, 1850, 266; Amelia Hadley, "Journal of Travails to Oregon," June 9, 1851, 74; Sawyer, "Overland to California," July 28, 1852, 24.

71. Egbert, "Mother's Diary," June 13, 1852; Babcock, "Journal of the Road," July 2, 1859; Breyfogle, *Diary of J. D. Breyfogle*, June 14, 20, 1849; Clayton, *Latter-day Saints' Emigrants' Guide*, 157–58, 168; J. Palmer, *Journal of Travels over the Rocky Mountains*, July 21, August 1, 8, 1845, 35, 36, 43; Johnson and Winter, *Route Across the Rocky Mountains*, 70; Batchelder, "Journal of a Tour Across the Continent," August 24, 1849; Shepherd, "Journal of Travel," July 4, 1850, 19.

72. Batchelder, "Journal of a Tour Across the Continent," July 19, 1849, 42; McLaughlin, *Sketch of a Trip*, 11; Johnson and Winter, *Route Across the Rocky Mountains*, 76.

73. J. Palmer, *Journal of Travels over the Rocky Mountains*, August 20, 1845, 46; Sawyer, "Overland to California," July 4, 5, 1852, 18; G. Belshaw, "Diary," August 11, 1853, 34; Egbert, "Mother's Diary," July 20, 1852; Judson, *Pioneer's Search for an Ideal Home*, 56; Johnson and Winter, *Route Across the Rocky Mountains*, 76.

74. Porter, "By Ox Team to California," 231; J. Clark, *Overland to the Gold Fields of California*, 245; J. Palmer, *Journal of Travels over the Rocky Mountains*, July 19, 1845, 34; Baldy, "Journal," June 23, 1852; "Life in Nebraska."

75. Smedley, *Across the Plains in '62*, June 2, 1862; Crockett to Friends, February 16, 1845; C. Smith, *Journal of a Trip to California*, May 19, 1850, 31; Porter, "By Ox Team to California," 227; Empey, "Journal of William A. Empey," 1847.

76. Foster to family, July 13, 1849, in *Foster Family*, 31; "The Prairie Dog"; Lienhard, "California Immediately Before and After the Discovery of Gold," 8.

77. Alexander, "Little Story," 4; B. Clark, "Diary of a Journey from Missouri to California," July 25, 1849; Hamilton, *Pioneer of Fifty-Three*, 40.

78. Arms, "Journal Describing Overland Journey," May 29, 1849.

79. Enos, *Across the Plains*, 1850; Baldy, "Journal," June 6, 1852; Barnett, *Long Trip in a Prairie Schooner*, 44.

80. Breyfogle, *Diary of J. D. Breyfogle*, May 12, 1849; O. Taylor Papers, "Diary," August 8, 1851 (see chap. 4, n. 57).

7. End of the Trail

1. Hine and Faragher, *American West*, 436; Nichols, "Across the Plains in 1844" (see chap. 4, n. 1); Enos, *Across the Plains*, letter 3.

2. Frizzell, *Across the Plains to California,* June 8, 1852, 23.

3. Lienhard, "California Immediately Before and After the Discovery of Gold," 146 (see chap. 3, n. 13).

4. Sawyer, "Overland to California," August 7, 1852, 28 (see chap. 3, n. 22); Hutchings, "Diaries," September 7, 1848 (see chap. 3, n. 20); Batchelder, "Journal of a Tour Across the Continent," July 23, September 21, 1849 (see chap. 3, n. 11); S. Richey to J. Richey, November 26, 1853, 4 (see chap. 4, n. 90); C. A. Tuttle to M. Tuttle, August 19, 1849 (see chap. 3, n. 11); Shepherd, "Journal of Travel," August 8, 1850 (see chap. 3, n. 8).

5. Todd Berens Papers, folder: Arthur Shearer Biography, *Hannibal Journal* excerpts, MS 491, Ex. 1, Fol. 2, UU; Mattes, *Great Platte River Road,* 15.

6. Smedley, *Across the Plains in '62,* August 4, 1862.

7. Porter, "By Ox Team to California," 196; J. Clark, *Overland to the Gold Fields of California in 1852,* August 31, 1852, 295; McCall, *Great California Trail,* September 17, 1849, 84; Smedley, *Across the Plains in '62,* June 28, 1862.

8. Barnett, *Long Trip in a Prairie Schooner,* 134.

9. Starr, "Diary," May 31, 1850, 10 (see chap. 4, n. 29).

Bibliography

Manuscript Collections

Bancroft Library, University of California–Berkeley
California Historical Society, San Francisco
Huntington Library, San Marino, CA
Merrill J. Mattes Research Library, Independence, MO
Multnomah County Central Library, Portland, OR
Oregon Historical Society, Portland
Public Record Office of Northern Ireland, Belfast
State Historical Society of Missouri, Columbia
University of Utah, Special Collections, Salt Lake City
Utah State Historical Society, Salt Lake City

Newspapers

Frontier Guardian (Kanesville, IA), 1849, 1850
Huron Reflector (Norwalk, OH), 1845
Nautical Standard and Steam Navigation Gazette (London), 1851
Sauk County Standard (Baraboo, WI), 1851
Southport Telegraph (Kenosha, WI), 1842
Watertown (WI) Chronicle, 1849

Other Sources

Abell, Mrs. L. G. *Woman in Her Various Relations: Containing Practical Rules for American Females.* New York: J. M. Fairchild, 1855.

Ackley, Mary E. *Crossing the Plains and Early Days in California: Memories of Girlhood Days in California's Golden Age.* San Francisco: privately printed, 1928.

Aird, Polly. "Bound for Zion: The Ten- and Thirteen-Pound Emigrating Companies, 1853–54." *Utah Historical Quarterly* 70, no. 4 (2002): 300–325.

Allen, O. *Allen's Guide Book and Map to the Gold Fields of Kansas and Nebraska and Great Salt Lake City.* Reprinted with notes by LeRoy R. Hafen. [Denver]: Nolie Mumey, 1953.

Anderson, Virginia DeJohn. *Creatures of Empire: How Domestic Animals Transformed Early America.* Oxford: Oxford University Press, 2004.

Ashton, John. "History of Jack Stock and Mules in Missouri" [August 1924]. *In Jack Stock and Mules in Missouri.* Rev. ed. Columbia: Curators of the University of Missouri, 1987.

Avery, Rev. J. H. *The Land of Ophir, Ideal and Real: A Discourse Delivered at Austinburg, Ohio, Before a Company About Proceeding to California.* New York: Edward O. Jenkins, 1853.

Barnett, Joel. *A Long Trip in a Prairie Schooner.* Whittier, CA: Western Stationery, 1928.

Beecher, Catherine E. *A Treatise on Domestic Economy.* Rev. ed. New York: Harper & Brothers, 1848.

Beers, Diane L. *For the Prevention of Cruelty: The History and Legacy of Animals Rights Activism in the United States.* Athens: Swallow Press / Ohio University Press, 2006.

Belshaw, Maria Parsons. *Crossing the Plains to Oregon in 1853.* Edited by Michael L. Tate. Fairfield, WA: Ye Galleon Press, 2000.

Bierer, Bert W. *A Short History of Veterinary Medicine in America.* [East Lansing]: Michigan State University Press, 1955.

Bigland, John. *A Natural History of Animals.* Philadelphia: Grigg & Elliot, 1844.

Billington, Ray Allen. "Books That Won the West." *American West* 4 (1967): 25–32, 72–75.

———. *The Far Western Frontier, 1830–1860.* Albuquerque: University of New Mexico Press, 1995.

Billington, Ray Allen, and Martin Ridge. *Westward Expansion: A History of the American Frontier.* 6th ed. Albuquerque: University of New Mexico Press, 2001.

Boorstin, Daniel J. *The Americans: The National Experience.* New York: Vintage Books, 1965.

Born, Fred J. *The History of the Horse Doctor (*Equarius medicus*): A Brief Summary of 4000 Years of Veterinary History* [CD-ROM]. Self-published, 2007.

Bradley, Melvin. *The Missouri Mule: His Origin and Times.* Vol. 1. Columbia: University of Missouri, 1993.

Breyfogle, J. D. *Diary of J. D. Breyfogle, Sr.: Covering His Experiences During His Overland Trip to California During the Gold Rush in 1849.* N.p., [189?].

Brown, Milton W. *American Art to 1900: Painting, Sculpture, Architecture.* New York: Harry N. Abrams, 1977.

Brown, William Richard. *An Authentic Wagon Train Journal of 1853: From Indiana to California.* Privately printed.

Buchanan, George W. "Letter and Circular of Information for Prospective Emigrants to Oregon." *Oregon Historical Quarterly* 11 (1910): 302–12.

Burrell, Mary. "Council Bluffs to California, 1854." In vol. 6 of *Covered Wagon Women: Diaries and Letters from the Western Trails, 1853–1854,* edited by Kenneth L. Holmes. Lincoln: University of Nebraska Press, 1986.

California Wolf Center. "Comparing Wolves and Coyotes." http://www.californiawolfcenter.org.

Cayton, Andrew R. L. "Lewis and Clark Expedition." In *Encyclopedia of the Enlightenment,* edited by Alan Charles Kors. New York: Oxford University Press, 2003.

Child, Andrew. *Overland Route to California.* Introduction by Lyle H. Wright. Los Angeles: N. A. Kovach, 1946.

Child, Lydia. *The Mother's Book* [1831]. http://digital.library.upenn.edu/women/child/book/book.htm.

Clark, Bennett. "Diary of a Journey from Missouri to California in 1849." Edited by Ralph P. Bieber. *Missouri Historical Review* 23, no. 1 (1928): 3–43.

Clark, John Hawkins. *Overland to the Gold Fields of California in 1852.* Edited by Louise Barry. Reprint, Topeka: Kansas State Printing Plant, 1942.

Clayton, William. *The Latter-day Saints' Emigrants' Guide: Being a Table of Distances. . .from Council Bluffs to the Valley of the Great Salt Lake.* St. Louis: Mo. Republican Steam Power Press Chambers & Knapp, 1848. In Leander V. Loomis, *A Journal of the Birmingham Emigrant Company,* edited by Edgar M. Ledyard. Salt Lake City: Legal Printing, 1928.

Cole, William L. *California: Its Scenery, Climate, Productions and Inhabitants.* New York: Irish-American Office, 1871.

Coleman, Jon T. *Vicious: Wolves and Men in America.* New Haven, CT: Yale University Press, 2004.

Cooke, Lucy Rutledge. "Letters on the Way to California." In vol. 4 of *Covered Wagon Women: Diaries and Letters from the Western Trails, 1852,* edited by Kenneth L. Holmes. Lincoln: University of Nebraska Press, 1985.

Cornaby, Hannah. *Autobiography and Poems.* Salt Lake City: J. C. Graham, 1881.

Crawford, Medorem. "Journal of Medorem Crawford." *Sources of the History of Oregon* 1, no. 1 (1897): 4–26.

Cronon, William. *Changes in the Land: Indians, Colonists, and the Ecology of New England.* New York: Hill and Wang, 1983.

Crosby, Alfred W., Jr. *The Columbian Exchange: Biological and Cultural Consequences of 1492.* Westport, CT: Greenwood Press, 1972.

Dadd, G. H. *The American Cattle Doctor.* New York: Orange Judd, 1851. Reprint, Bedford, MA: Applewood Books, n.d.

———. "Our Platform." *American Veterinary Journal* 1, no. 1 (1855): 1–2.

Dary, David A. "Buffalo." In *The New Encyclopedia of the American West,* edited by Howard R. Lamar. New Haven, CT: Yale University Press, 1998.

———. *The Buffalo Book: The Full Saga of the American Animal.* Athens: Swallow Press/Ohio University Press, 1989.

———. *The Oregon Trail: An American Saga.* New York: Alfred A. Knopf, 2004.

Davidson, Marshall B. *The American Heritage History of the Artists' America.* New York: American Heritage, 1973.

Delano, Alonzo. *California Correspondence.* Sacramento: Sacramento Book Collectors Club, 1952.

Dodds, Gordon B. "Fort Bridger." In *The New Encyclopedia of the American West,* edited by Howard R. Lamar. New Haven, CT: Yale University Press, 1998.

———. "Fort Hall." In *The New Encyclopedia of the American West,* edited by Howard R. Lamar. New Haven, CT: Yale University Press, 1998.

Dodds, Gordon B., and Richard N. Ellis. "Fort Laramie." In *The New Encyclopedia of the American West,* edited by Howard R. Lamar. New Haven, CT: Yale University Press, 1998.

Dolan, Edward F., Jr. *Animal Rights.* New York: Franklin Watts, 1986.

"Effects of Opium." *American Veterinary Journal* 14 (November 1956): 432–33.

Elliott, Russell R., and William D. Rowley. *History of Nevada.* 2nd ed. Lincoln: University of Nebraska Press, 1987.

Ellis, Henrietta Catherine, and James Addison Ellis. *Across the West by Covered Wagon, 1865.* Edited by Beryl Leerick. Honolulu: Hawaii State Newspapers, 1961.

Ellis, Richard N. "Fort Kearny." In *The New Encyclopedia of the American West,* edited by Howard R. Lamar. New Haven, CT: Yale University Press, 1998.

Empey, William A. "The Journal of William A. Empey, 1847." Edited by Dale L. Morgan. *Annals of Wyoming* 21 (1949).

Enos, A. A. *Across the Plains in 1850.* [Stanton, NE?]: n.p., 1905.

Faragher, John Mack. *Women and Men on the Overland Trail.* 2nd ed. New Haven, CT: Yale Nota Bene 2001.

Flint, Dr. Thomas. *Diary of Dr. Thomas Flint, California to Maine and Return, 1851–55.* Los Angeles: Historical Society of Southern California, 1923.

Flores, Dan. "Bison Ecology and Bison Diplomacy: The Southern Plains from 1800 to 1850." *Journal of American History* 78, no. 2 (1991): 465–85.

Foster, Isaac. *The Foster Family: California Pioneers.* Santa Barbara, CA: Schauer Printing Studio, 1925.

Francione, Gary L., and Anna E. Charlton. "Animal Advocacy in the 21st Century: The Abolition of the Property Status of Nonhumans." In *Animal Law and the Courts: A Reader.* St. Paul, MN: Thomson West, 2008.

Franklin, Benjamin. *Poor Richard's Almanack.* Waterloo, IA: USC, 1914.

Frink, Margaret A. "Adventures of a Party of Gold-Seekers." In vol. 2 of *Covered Wagon Woman: Diaries and Letters from the Western Trails, 1850,* edited by Kenneth L. Holmes. Lincoln: University of Nebraska Press, 1983.

Frizzell, Lodisa. *Across the Plains to California in 1852.* New York: New York Public Library, 1915.

Frush, Wm. H. *The Overland Diary, May 15–September 25, 1850.* Reprint, Fairfield, WA: Ye Galleon Press, 2000.

Fry, Frederich. *Fry's Traveler's Guide and Descriptive Journal of the Great North-Western Territories.* Edited by Arrell Morgan Gibson. 2nd ed. Reprint, Oklahoma City: Dean Press, 1971.

Fuller, Emeline J. "Left by the Indians." In vol. 96 of *Narratives of Indian Captivities.* New York: Garland, 1978.

Goltra, Elizabeth Julia. *Journal Kept by Mrs. E. J. Goltra of Her Travels Across the Plains in the Year 1853.* Eugene, OR: Lane County Historical Society, 1970.

Goodwin, Joan. "Lydia Maria Child." In *Dictionary of Unitarian & Universalist Biography.* http://www25.uua.org/uuhs/duub/articles/lydiamariachild.html.

Graves, Russell A. *The Prairie Dog: Sentinel of the Plains.* Lubbock: Texas Tech University Press, 2001.

Greeley, Horace. *An Overland Journey from New York to San Francisco in the Summer of 1859.* Edited by Charles T. Duncan. Reprint, New York: Alfred A. Knopf, 1964.

Greene, Ann Norton. *Horses at Work: Harnessing Power in Industrial America.* Cambridge, MA: Harvard University Press, 2008.

Grier, Katherine C. "'The Eden of Home': Changing Understandings of Cruelty and Kindness to Animals in Middle-Class American Households, 1820–1900." In *Animals in Human Histories: The Mirror of Nature and Culture.* Rochester, NY: University of Rochester Press, 2002.

———. "'Kindness to All Around': The Changing Ethics of Animal Treatment in the Middle-Class Household, 1820–1870." *Between the Species* (Fall 1992): 181–93.

———. *Pets in America: A History.* Chapel Hill: University of North Carolina Press, 2006.

Groseclose, Barbara. *Nineteenth-Century American Art.* Oxford: Oxford University Press, 2000.

Hadley, Amelia. "Journal of Travails to Oregon." In vol. 3 of *Covered Wagon Women: Diaries and Letters from the Western Trails, 1851,* edited by Kenneth L. Holmes. Lincoln: University of Nebraska Press, 1984.

Hamilton, Sarah Watson. *A Pioneer of Fifty-Three.* Albany, OR: Herald Press, 1905.

Hammer, Jacob. *This Emigrating Company: The 1844 Oregon Trail Journal of Jacob Hammer.* Edited by Thomas A. Rumer. Spokane, WA: Arthur H. Clark, 1990.

Hastings, Lansford W. *The Emigrants' Guide to Oregon and California.* 1845. Reprint, Bedford, MA: Applewood Books, 1994.

Hauer, John. *The Natural Superiority of Mules.* Guilford, CT: Lyons Press, 2005.

Haverstock, Mary Sayre. "An American Bestiary." *Art in America* 58, no. 4 (1970): 38–71.

———. *An American Bestiary.* New York: Harry N. Abrams, 1979.

Hayes, Derek. *Historical Atlas of the American West with Original Maps.* Berkeley: University of California Press, 2009.

Hester, Sallie. "The Diary of a Pioneer Girl." In vol. 1 of *Covered Wagon Women: Diaries and Letters from the Western Trails, 1840–1849,* edited by Kenneth L. Holmes. Lincoln: University of Nebraska Press, 1983.

Hine, Robert V., and John Mack Faragher. *The American West: A New Interpretive History.* New Haven, CT: Yale University Press, 2000.

Homer. *The Iliad.* Translated by Samuel Butler. http://www.literaturepage.com.

Horn, Hosea B. *Horn's Overland Guide to California.* New York: J. H. Colton, 1853.

Horsman, Reginald. *Race and Manifest Destiny: The Origins of American Racial Anglo-Saxonism.* Cambridge, MA: Harvard University Press, 1981.

Howe, Daniel Walker. "American Victorianism as a Culture." *American Quarterly* 27, no. 5 (1975): 507–32.

Hulse, James W. *The Silver State: Nevada's Heritage Reinterpreted.* Reno: University of Nevada Press, 1991.

Hunt, Nancy A. "By Ox-Team to California." *Overland Monthly* 67 (1916): 317–26.

Hygnstrom, Scott E. "Black-Tailed Prairie Dog." In *Encyclopedia of the Great Plains,* edited by David J. Wishart. Lincoln: University of Nebraska Press, 2004.

"Importance of Veterinary Science." *American Veterinary Journal* 2 (1857): 155.

Irving, Washington. *A Tour of the Prairies.* Edited by Joseph B. Thoburn and George C. Wells. Oklahoma City: Harlow, 1930.

Ismay, Dr. John. "Selenium Poisoning in Horses (Alkali Disease)." 2012. http://igrow.org/livestock/horse/selenium-poisoning-in-horses-alkali-disease/.

Johnson, Overton, and William H. Winter. *Route Across the Rocky Mountains.* Edited by Angela Firkus. 1846. Reprint, West Lafayette, IN: NotaBell Books, 2000.

Jones, Gerald E. "The Gospel and Animals." August 1972. http://www.lds.org/ensign/print/1972/08/the-gospel-and-animals?lang=eng.

———. "Zoophily in Mormonism." *Restoration* 4, no. 4 (1985): 1, 7–10.

Jones, Gerald E., and Scott S. Smith. *Animals and the Gospel: A History of Latter-day Saint Doctrine on the Animal Kingdom.* Thousand Oaks, CA: Millennial, 1980.

Judson, Phoebe Goodell. *A Pioneer's Search for an Ideal Home.* Lincoln: University of Nebraska Press, 1984.

Kalof, Linda. *Looking at Animals in Human History.* London: Reaktion Books, 2007.

Katcher, Aaron Honori, and Alan M. Beck. "Animal Companions: More Companion than Animal." In *Man & Beast Revisited,* edited by Michael H. Robinson and Lionel Tiger. Washington, DC: Smithsonian Institution Press, 1991.

Kirkpatrick, Thomas Jefferson. *The Kirkpatrick Story: The Day by Day Report of the Trek Across the Plains by the Kirkpatrick Party in 1854.* Reprint, Orland, CA: Orland Register, 1954.

Klaus, Michael W. "Mormon Cricket." Tree Fruit Research and Extension Center. Orchard Pest Management Online, Washington State University, January 2008. http://jenny.tfrec.wsu.edu/opm.displaySpecies.php?pn=253.

Knight, Amelia. "Iowa to the Columbia River." In vol. 6 of *Covered Wagon Women: Diaries and Letters from the Western Trails, 1853–1854,* edited by Kenneth L. Holmes. Lincoln: University of Nebraska Press, 1986.

Koford, Carl B. "Prairie Dogs." In *The New Encyclopedia of the American West,* edited by Howard R. Lamar. New Haven, CT: Yale University Press, 1998.

Kroll, Helen B. "The Books That Enlightened the Emigrants." *Oregon Historical Quarterly* 45, no. 2 (1944): 103–23.

Larkin, Jack. *The Reshaping of Everyday Life, 1790–1840.* New York: Harper & Row, 1988.

Lewis, Capt. John I. *My Garden of Roses; or, The Footnote of Life.* [Berkeley, CA, 1913?].

Loomis, Leander V. *A Journal of the Birmingham Emigrating Company: The Record of a Trip from Birmingham, Iowa, to Sacramento, California, in 1850.* Edited by Edgar M. Ledyard. Salt Lake City: Legal Printing, 1928.

Loughary, Harriet A. "Travel and Incidents, 1864." In vol. 8 of *Covered Wagon Women: Diaries and Letters from the Western Trails, 1862–1865,* edited by Kenneth L. Holmes. Lincoln: University of Nebraska Press, 1989.

Loveland, Cyrus C. *California Trail Herd: The 1850 Missouri-to-California Journal of Cyrus C. Loveland.* Edited by Richard H. Dillon. Los Gatos, CA: Talisman Press, 1961.

Lubin, David M. "Art in an Age of National Expansion: Genre and Landscape Painting." In *Art in America: 300 Years of Innovation.* London: Merrell, 2007.

Marcy, Capt. Randolph B. *The Prairie Traveler: A Hand-Book for Overland Expeditions.* 1859. Reprint, Bedford, MA: Applewood Books, 1993.

Mattes, Merrill J. *The Great Platte River Road: The Covered Wagon Mainline via Fort Kearny to Fort Laramie.* 1969. Reprint, Lincoln: University of Nebraska Press, 1987.

McCall, A. J. *The Great California Trail in 1849.* Bath, NY: Steuben Courier, 1882.

McCandless, Perry. *A History of Missouri.* Vol. 2, *1820 to 1860.* Columbia: University of Missouri Press, 2000.

McLaughlin, Daniel. *Sketch of a Trip from Omaha to Salmon River.* Chicago: Gordon Martin, 1954.

McShane, Clay, and Joel A. Tarr. *The Horse in the City: Living Machines in the Nineteenth Century.* Baltimore: Johns Hopkins University Press, 2007.

Menache, Sophia. "Animals in the Bible and Qur'an." In *The Encyclopedia of Religion and Nature.* London: Thoemmes Continuum, 2005.

Mothershead, Harmon. "River Town Rivalry for the Overland Trade." *Overland Journal* 7, no. 2 (1989): 14–23.

Moynihan, Ruth Barnes. "Children and Young People on the Overland Trail." *Western Historical Quarterly* 6, no. 3 (1975): 279–95.

Mullan, Capt. John. *Miners and Travelers' Guide to Oregon, Washington, Idaho, Montana, Wyoming, and Colorado via the Missouri and Columbia Rivers.* New York: Wm. M. Franklin, 1865.

Munkres, Robert L. "Wagon Train Animals." *Wyoming Annals* 65 (Summer–Fall 1993): 15–27.

Murphy, Virginia Reed. *Across the Plains in the Donner Party: A Personal Narrative of the Overland Trip to California.* Silverthorne, CO: Vistabooks, 1995.

Myres, Sandra L. *Western Women and the Frontier Experience, 1800–1915.* Albuquerque: University of New Mexico Press, 1982.

Nagel, Paul C. *Missouri: A Bicentennial History.* New York: W. W. Norton, 1977.

National Geographic. "Grizzly Bear." http://animals.nationalgeographic.com/animals/mammals/grizzly-bear/.

Osgood, Ernest S. "Our Dog Scannon: Partner in Discovery." *Montana: The Magazine of Western History* 26, no. 3 (1976): 8–17.

O'Sullivan, John L. "Annexation." *United States Magazine and Democratic Review* 17, no. 85 (1845): 5–6.

Paine, Thomas. *The Writings of Thomas Paine: The Age of Reason, Part I.* Edited by Moncure Daniel Conway. http://www.gutenberg.org/files/3743/3743-h/3743-h.html.

Palmer, Joel. *Journal of Travels over the Rocky Mountains.* 1847. Reprint, Ann Arbor, MI: University Microfilms, 1966.

Palmer, Rosemary Gudmundson. *Children's Voices from the Trail: Narratives of the Platte River Road.* Spokane, WA: Arthur H. Clark, 2002.

Parkman, Francis. *The Oregon Trail.* 1849. Reprint, New York: New American Library, 1950.

Parrish, William E., Charles T. Jones Jr., and Lawrence O. Christensen. *Missouri: The Heart of the Nation.* 2nd ed. Arlington Heights, IL: Harlan Davidson, 1992.

Parsons, Lucena. "An Overland Honeymoon." In vol. 2 of *Covered Wagon Women: Diaries and Letters from the Western Trains, 1850,* edited by Kenneth L. Holmes. Lincoln: University of Nebraska Press, 1983.

Peters, Arthur King. *Seven Trails West.* New York: Abbeville Press, 1996.

Phelps, Norm. *The Longest Struggle: Animals Advocacy from Pythagoras to Peta.* New York: Lantern Books, 2007.

Platt, Philip L., and N. Slater. *Travelers' Guide Across the Plains upon the Overland Route to California.* Reprint, San Francisco: John Howell-Books, 1963.

Poetry Foundation. "Lydia Huntley Sigourney." http://www.poetryfoundation.org/bio/lydia-huntley-sigourney.

Popplewell, Frank S. "St. Joseph, Missouri, as a Center of the Cattle Trade." *Missouri Historical Review* 32, no. 4 (1938): 443–57.

Porter, Lavina Honeyman. "By Ox Team to California." In *Westward Journeys: Memoirs of Jesse A. Applegate and Lavinia Honeyman Porter Who Traveled the Overland Trail,* edited by Martin Ridge. Chicago: Lakeside Press, 1989.

Powers, Mary Rockwood. *A Woman's Overland Journal to California.* Edited by W. B. Thorsen. Fairfield, WA: Ye Galleon Press, 1985.

Preece, Rod. *Brute Souls, Happy Beasts, and Evolution: The Historical Status of Animals.* Vancouver: UBC Press, 2005.

Primatt, Humphry. *A Dissertation on the Duty of Mercy and Sin of Cruelty to Brute Animals.* London: R. Hett, 1776.

Pringle, Laurence. *The Animal Rights Controversy.* San Diego: Harcourt Brace Jovanovich, 1989.

Rachels, James. "Animals and Ethics." In *Routledge Encyclopedia of Philosophy.* London: Routledge, 1998.

Reid, John Phillip. *Law for the Elephant: Property and Social Behavior on the Overland Trail.* San Marino, CA: Huntington Library, 1997.

Reynolds, David A. *Waking Giant: America in the Age of Jackson.* New York: Harper Perennial, 2008.

Ringo, Mary. "The 1864 Journal." In vol. 8 of *Covered Wagon Women: Diaries and Letters from the Western Trails, 1862–1865,* edited by Kenneth L. Holmes. Lincoln: University of Nebraska Press, 1989.

Rogers, Katharine M. *First Friend: A History of Dogs and Humans.* New York: St. Martin's Press, 1980.

Royce, Sarah. *Across the Plains: Sarah Royce's Western Narrative.* Edited by Jennifer Dawes Adkison. Tucson: University of Arizona Press, 2009.

Russow, Lilly-Marlene. "Changing Perceptions of Animals: A Philosophical View." In *Perceptions of Animals in American Culture,* edited by R. J. Hoage. Washington, DC: Smithsonian Institution Press, 1989.

Schlesinger, Arthur M. *Learning How to Behave: A Historical Study of American Etiquette Books.* New York: Cooper Square, 1968.

Schwartz, Marion. *A History of Dogs in the Early Americas.* New Haven, CT: Yale University Press, 1997.

Scott, Sir Walter. "The Lady of the Lake." Edited by William J. Rolfe. N.p., 1883. http://www.online-literature.com/walter_scott/2561/.

Searls, Niles. *The Diary of a Pioneer, and Other Papers.* San Francisco: Pernau-Walsh, 1940.

Sears, Clara Endicott. *Highlights Among the Hudson River Artists.* Boston: Houghton Mifflin, 1947.

Serpell, James. *In the Company of Animals: A Study of Human-Animal Relationships.* Oxford: Basil Blackwell, 1986.

Shively, J. M. *Route and Distances to Oregon and California.* Vol. 2. Reprint, Lincoln: University of Nebraska Press, 1963.

Sigourney, Lydia H. *Letters to Mothers.* 6th ed. New York: Harper & Brothers, 1845.

Smedley, William. *Across the Plains in '62.* Denver: n.p., 1916.

Smith, C. W. *Journal of a Trip to California, Across the Continent from Weston, Mo., to Weber Creek, Cal., in the Summer of 1850.* Edited by R. W. G. Vail. New York: Cadmus Book Shop, 1920.

Smith, Donna Campbell. *The Book of Mules: Selecting, Breeding, and Caring for Equine Hybrids.* Guilford, CT: Lyons Press, 2009.

Smith, Emily. "Wolves in the Wild." *Allanimals* 16, no. 5 (2014): 26–31.

Smithcors, J. F. *The Veterinarian in America, 1625–1975.* Santa Barbara, CA: American Veterinary Publications, 1975.

Spence, Clark C. "Canines to Canaan." *American Heritage* 32, no. 2 (1981): 58–64.

———. "Pioneers with Wagging Tails: Dogs on the Trail to Oregon." *Idaho Yesterdays* 25, no. 2 (1981): 27–31.

Stabaek, Tosten Kittelsen. "An Account of a Journey to California in 1852." *Studies and Records* 4 (1929): 99–124.

Tarbell, Dr. J. *The Emigrant's Guide to California.* Keokuk, IA: Whig Book and Job Office, 1853.

Taylor, Rachel. "Overland Trip Across the Plains." In vol. 6 of *Covered Wagon Women: Diaries and Letters from the Western Trails, 1853–1854,* edited by Kenneth L. Holmes. Lincoln: University of Nebraska Press, 1986.

Teigen, Philip M. "Nineteenth-Century Veterinary Medicine as an Urban Profession." *Veterinary Heritage* 23, no. 1 (2000).

Tetherow, Solomon. *The Organizational Journal of an Emigrant Train of 1845*. Eugene, OR: Lane County Pioneer Historical Society, 1960.

Thissell, G. W. *Crossing the Plains in '49*. Oakland, CA: n.p., 1903.

Tonna, Charlotte E. B. *Kindness to Animals; or, The Sin of Cruelty Exposed and Rebuked*. Philadelphia: American Sunday-School Union, 1845. http://msupress.msu.edu/bookTemplate.php?bookID=2163.

Troxel, Kathryn. "Food of the Overland Emigrants." *Oregon Historical Quarterly* 56, no. 1 (1955): 12–26.

Turner, James. *Reckoning with the Beast: Animals, Pain, and Humanity in the Victorian Mind*. Baltimore: Johns Hopkins University Press, 1980.

Unruh, John D., Jr. *The Plains Across: The Overland Emigrants and the Trans-Mississippi West, 1840–60*. Urbana: University of Illinois Press, 1993.

Waldau, Paul. "Animals." In *Encyclopedia of Religion*. 2nd ed. Detroit: Thomon-Gale, 2005.

Ware, Joseph E. *The Emigrants' Guide to California*. 1849. Reprint, Princeton, NJ: Princeton University Press, 1932.

Weinberg, H. Barbara, and Carrie Rebora Barratt. "American Scenes of Everyday Life, 1840–1910." In *Heilbrunn Timeline of Art History*. New York: Metropolitan Museum of Art, 2000. http://www.metmuseum.org/toah/hd/scen/hd_scen.htm.

Wells, Eugene T. "The Growth of Independence, Missouri, 1827–1850." *Bulletin of the Missouri Historical Society* 16 (October 1959): 33–46.

White, Richard. *"It's Your Misfortune and None of My Own": A New History of the American West*. Norman: University of Oklahoma Press, 1991.

Williams, Jacqueline. *Wagon Wheel Kitchens: Food on the Oregon Trail*. Lawrence: University Press of Kansas, 1993.

Willoughby, Robert J. *Robidoux's Town: A Nineteenth-Century History of St. Joseph, Missouri*. St. Joseph, MO: St. Joseph Museums, 2006.

Wood, Elizabeth. "Journal of a Trip to Oregon." In vol. 3 of *Covered Wagon Women: Diaries and Letters from the Western Trails, 1851*, edited by Kenneth L. Holmes. Lincoln: University of Nebraska Press, 1984.

Wyeth, John B. *Oregon*. Cambridge: John B. Wyeth, 1833. Reprint, Ann Arbor, MI: University Microfilms, 1966.

Youatt, William. *The Horse*. Philadelphia: Porter & Coates, 1843.

Zellman, Michael David, comp. *American Art Analog*. Vol. 1, *1688–1842*. New York: Chelsea House, 1986.

Zieber, Eugenia. "Journal of Our Journey to Oregon." In vol. 3 of *Covered Wagon Women: Diaries and Letters from the Western Trails*, edited by Kenneth L. Holmes. Lincoln: University of Nebraska Press, 1984.

Index